GENERATORS AND INVERTERS
BUILDING SMALL COMBINED HEAT AND POWER SYSTEMS FOR REMOTE LOCATIONS AND EMERGENCY SITUATIONS

STEPHEN D. CHASTAIN

B.Sc. MECHANICAL ENGINEERING AND MATERIALS SCIENCE
UNIVERSITY OF CENTRAL FLORIDA

GENERATORS AND INVERTERS
BUILDING SMALL COMBINED HEAT AND POWER SYSTEMS FOR REMOTE LOCATIONS AND EMERGENCY SITUATIONS

By Stephen D. Chastain

Copyright© 2006 By Stephen D. Chastain

Jacksonville, FL All Rights Reserved

Printed in USA

ISBN 978-0-9702203-5-6

Titles by Stephen Chastain:

Generators and Inverters: Building Small Combined Heat and Power Systems for Remote Locations and Emergency Situations

The Small foundry Series:

Volume I.   Iron Melting Cupola Furnaces for the Small Foundry

Volume II. Build an Oil-Fired Tilting Furnace

Volume III. Metal Casting: A Sand Casting Manual Vol. I

Volume IV. Metal Casting: A Sand Casting Manual Vol. II

Volume V   Making Pistons for Experimental and Restoration Engines

Web Site: http://**StephenChastain.Com**   stevechastain@hotmail.com

TABLE OF CONTENTS

ACKNOWLEDGEMENTS:

No project can be completed without input from others. Often, seemingly unrelated events play a role in its successful completion. Several people contributed in many ways to this project. I would first like to thank **Liu Xinping** of Weichai Power for her world class service as a supplier of diesel engines and parts. Any company would be *lucky* to have her on their staff. I would like to thank **Ken Hyde** of Florida Community College for his expertise in brazing and welding the heat exchanger tubing. **Chris Chan-A-Sue** of **Modine** supplied raw test data for various heat exchanger configurations. The people at **Chevron** supplied much data regarding physical properties of various oils. **Norm Lillybeck**, the engineer's engineer, has an eye for the smallest details and helps keep me out of trouble. **George Hapsis** of **Zephyr Audio** provided a frequency analyzer. **Richard Nunn**, our local weather authority, checks my lightening data. Thanks to **Scott Merrell** for the new cover design. **Andre Ancelin** and **Larry Wallen** provided critical input at a few points in the project. **Jackie Dostie** always brightens the room with her cheerful eyes and smile. **Jim Thompson's** quick wit keeps the drama of daily life in perspective. And thanks to my friend **Maria** for supplying some of the drama.

INTRODUCTION:

As a young boy, I was fascinated by power generation. After seeing a diagram of a DC generator, I ripped one of my toys apart removing the permanent magnet motor. Attaching a small lamp from a string of Christmas tree lights, I would spin the motor's shaft and light up the bulb, being quite delighted with my discovery.

Several years later, I recovered an antique flathead motor from a riverbank. After setting up a small home foundry, I managed to produce almost all of the replacement parts needed to get the motor running. The natural progression of the project was to attach it to a generator and power my house. Florida storms frequently take out our power for 4 to 12 hours making the remotely started generator plant a real convenience.

During the summer of 2004, 4 major hurricanes struck the southeastern United States leaving millions without power. Tens of thousands were without power and water for several weeks. All refrigerated food had to be discarded. Many found the heat, even at night unbearable, and could not sleep. Surrounded by all of this misery, my little generator set supplied power for 3 families who lived in relative comfort, taking hot showers, using window air-conditioners and watching satellite TV.

This book is my response to both the "generator envy" and many questions I received during the ordeal. Here, I detail the development of my experimental "off the grid" power plant. While many standby generator sets can be pretty much assembly projects, calculations are provided so that you may see how and why I made my design decisions. The math is not particularly difficult and may be performed with an $8.95 scientific calculator available at Wal-Mart.

PURPOSE:

The purpose of this book is to describe and demonstrate the design and construction of a small experimental power plant to provide heat and power for remote locations. Radiator cooled standby units for emergency power are also described.

A 15 kW power plant will be built in a home shop and foundry from commonly available materials. A $49 scrap auto engine modified to run on propane or natural gas will power the generator. Combustion, the use of sewer gas and diesel power will also described.

To increase the efficiency of the plant, approximately 120,000 Btu per hour will be recovered from both the engine's water-jacket and exhaust. The recovered heat may be used for hot water production, radiant floor heating or pool heating.

To augment the plant, a commercial battery-inverter system is assembled. The inverter limits the plant's run time and keeps it fully loaded so that it operates at its highest efficiency.

Common switchgear is described for emergency power applications. Suggested minimum lightening protection, vibration isolation and noise control are also described.

Because the project is experimental and evolving, several solutions to a problem may be offered. It is up to the reader to judge which solution may be the best for their situation. Sufficient background material is provided so that the reader may understand the advantages and disadvantages of a particular solution.

Ammonia absorption refrigeration systems typically use waste heat to produce ice. It is possible that a small absorption chiller could be built to cool your house. These chillers were common between 1860 to the 1940's. The cycle is still used today, however on an industrial scale.

1. SAFETY:

ELECTROCUTION:

If not properly designed and installed and grounded, electrical systems can be hazardous. While you are familiar with the usual amperage ratings, unsafe currents are quite small. The lowest level of current that is perceptible is about 100 µA (micro-amp 1/1,000,000 amp) if it is concentrated at a sharp point. A large area, such as a handle, may require 1 mA (milli-amp 1/1000 amp) to be noticed.

Muscle contraction and pain may develop between 1 – 5 mA and is certain at 10 mA. Uncontrolled contraction, the inability to let go of an energized conductor can start at 6 mA. Up to 30 mA, the contractions increase in intensity and respiratory paralysis can occur. Above 30 mA, possibility of death from a number of causes, including cardiac fibrillation increase. Cardiac fibrillation is a condition in which the heart muscles contract in a random manner so that little blood is pumped.

Body tissue has a specific resistance. As current flows through a resistance, it generates heat. Current flowing through body tissue may cause burning, coagulation of blood and nerve damage. Dry, calloused skin can have an impedance approaching 1,000,000 ohms, which serves as a barrier to leakage currents. Damp or sweaty skin can provide a low impedance path for currents through body tissue.

LIGHTENING PROTECTION:

THUNDERCLOUDS AND LIGHTENING:

A thundercloud is a mass of electrical energy of perhaps 1000 coulombs (6.28 x 10^{18} electrons) or more distributed over a space as large as 30 cubic miles. Storm-cells forms from small cumulous clouds which are the tops of rising columns of air. If the convective currents are strong enough, these clouds grow upward at rates of approximately 100 feet per second or more. As the tops of the clouds reach approximately 28,000 feet, lightening begins to flash. A typical storm-cell will reach a height of 37,000 feet but may reach as high as 60,000 feet. The bottom of a cell is typically about 5000 to 9000 feet above the ground. The temperature at the bottom of a cell is usually about 60°F while the temperature at the top may be as low as -60°F. The diameter of a cell ranges from 300 feet to 30,000 feet. The lifetime for a typical cell is from 30 to 60 minutes.

One theory of cloud electrification states that as the rain drops fall through the cloud, they take on a tear drop shape. The bottoms of the drops take on a negative charge while the tops carry a positive charge. As the velocity of the drop increases relative to the rising air, the top of the teardrops break off into very small droplets carrying the positive charge upwards.

Thunder is caused when an electric spark traverses the air. A column of air, often several miles long, is suddenly heated and considerably expands creating the clap of thunder. Lightening travels at a speed of 186,000 miles per second while sound travels at 1090 feet per second or about 1 mile for every 5 seconds. The distance of the lightening may be estimated from the time until the thunder. The length of the rumble is relative to the length of the lightening bolt. A typical storm will have 480 to 720 lightening flashes with about 70% being cloud to cloud flashes or 120 to 240 being ground to cloud flashes. A typical lightening flash is about 25,000 amps and 30,000,000 volts in 4 pulses with a duration of about .4 seconds.

GROUNDING:

A lightening strike may damage a small powerplant or any connected equipment. Typically, the minimum grounding specifications of the electric code may prevent a fire but are inadequate for proper lightening protection of sensitive equipment.

Good grounding systems are largely dependent upon the resistance of the soil, resistance of the junctions in the grounding system and the capacity of the grounding conductors.

A grounding system based on ½-inch grounding rods driven 10 feet deep may well protect a structure in Boston, where the soil has a resistance of ½ ohm at 10 feet. But the same grounding system in Worchester where the soil's resistance is 3,000-ohms at 10 feet would offer little protection.

Soil resistance is affected by five factors that include (1) Type of Soil, (2) Depth, (3) Moisture Content (4) Temperature and (5) Transfer Resistance of the grounding media (rods, ground plates buried cables, mesh, etc.).

Rock will degrade the grounding ability relative to the amount of rock present in the soil.

Type of Soil	Resistance in Ohms		
	Average	Minimum	Maximum
Grounds containing fills such as cinders, ash or brine	14	3.5	41
Clay, shale, adobe, and slightly sandy loam with no stones or gravel	21	2	98
Clay, and loam with varying amounts of stones or gravel	93	6	800
sand, stones or gravel	554	35	2,700
	*adapted from Frydenlund, M		

The first few feet of soil are the driest and therefore have the highest resistance. Depth of the grounding rods greatly affects the resistance of the system for example; a grounding rod driven 10 feet deep may have a resistance of 100 ohms while a 30-foot rod in the same soil may have a resistance of only 4 ohms. Here in Florida, it is not uncommon to drive electrodes 50 feet deep. Resistance is determined by driving reference an electrode 300 feet from the grounding rod and checking the resistance between them. A second electrode 90° from the first may be used if the reading is unusually high. The National Electric Code recommends a maximum resistance of 25 ohms, however lightening protection specialist recommend a maximum of 5-ohms.

Moisture greatly increases the conductivity of the soil however moisture alone does not guarantee a low resistance path.

Soil resistance increases with lower temperatures and greatly increases in frozen soil.

Transfer Resistance refers to the resistance between the grounding media or rod and the surrounding soil. To decrease the transfer resistance, rods may be driven into a hole filled with scrap iron, charcoal or other conductive material called a *grounding reservoir*. These are usually located at a low spot where moisture collects. Where rock prevents driving of rods, extended cables, typically at least 24 feet long may be buried in trenches with conductive material.

Driven rods are available from 8-foot to 20-foot lengths and are made from copper, galvanized steel, or stainless steel. However copper is preferred to prevent corrosion of dissimilar metals Multiple rods are commonly used to lower the resistance. Distance between driven rods is at least equal to the depth of the rod. Sectional rods are used for deep grounding.

Connections are made with 4/0 copper cable and the junctions are brazed to prevent corrosion and the corresponding increase in resistance. A 4-inch radius is maintained of all bends.

VOLTAGE DROP AT A DISTANCE FROM THE GROUNDING ROD:

The earth is a poor conductor with a resistance of approximately 1 billion times that of copper. A ¾-inch diameter rod driven 8 feet into the ground may have a resistance of 25 ohms. The resistance may be viewed as a series of concentric circular cylinders around the grounding rod with the innermost cylinder next to the rod having the highest resistance. The resistance decreases rapidly with the distance from the rod as the surface area of each cylinder increases. Half of the 25-ohm resistance is likely to be contained inside of a 5-foot diameter cylinder around the rod. Therefore half of the voltage drop from current flowing through the rod would be within a 2-½ foot radius around the rod. If a typical lightening flash is 25,000 amps, the voltage at the rod relative to the earth would be 25 ohms x 25,000 amps = 625,000 volts. Half of this voltage, 312,500 volts would appear as a voltage drop between the rod and the earth at 2 ½ feet from the rod.

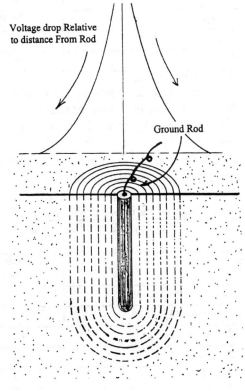

Voltage drop Relative to distance From Rod

Ground Rod

If a person was standing 15 feet away from the rod with their feet 3 feet apart (one foot at 15 feet from the rod and the other 18 feet from the rod), the voltage across the person's body would be equal to the voltage drop across the ground between his feet. This would at least give him a good jolt and may be fatal. While a person standing with both feet 15 feet away from the rod would have little potential difference across their body. This difference in "step voltage" explains why a horse may be killed by a lightening strike while a dog standing next to him may not.

2. EFFICIENCY OF ENERGY CONVERSION: Converting energy from some type of fuel such as coal, gas or oil into electricity and finally to work around your house is not a particularly efficient affair. For a typical utility, only about 35% of the energy released by burning the fuel actually reaches your house. The rest is lost as heat in the exhaust of the plant, heating of the wires, buss bars, transformers and the transmission lines. Ultimately all of the energy released as heat in the boiler reappears as heat or some other type of energy.

Efficiency of Energy Conversion

	Per Cent Efficiency	Per Cent of Energy Available
Coal		100
Boiler	85	85
Turbine	50	42.5
Generator	98	41.65
Transmission system	85	35.4
Pole Transformers	99	35

(Approximate efficiencies)

Engine generator sets are in a similar situation. Only about 25% to 30% of the heating value of the fuel is converted to mechanical energy and of that only about 85% is converted to electrical energy in the generator.

Conversion of the electrical energy back into light or mechanical energy is also an inefficient process. Therefore a small increase in the efficiency of an appliance results in a large savings of fuel. The efficiencies of various devices are listed below.

Efficiency of Energy Conversion

	Per Cent Efficiency	Per Cent of Energy Converted
Electricity into House		35
Incandescent Lamps	3.5	1.225
Fluorescent Lamps	20	7
Large Motors	92	32.2
Small Motors	70	24.5

3. WIND POWER: To know how much power a wind system will produce in a given location, you must know the output of the generator relative to the wind speed and you must have complete wind speed data for the site. Wind speed data is typically plotted as a graph for at least 1 to 2 years. Anything below 6 mph is discarded because the generator output is negligible. In the absence of good data, you can estimate from the average yearly winds at the site. The Weather Bureau records wind speeds hourly at stations all over the country.

Average Monthly Output in Kilowatt-Hours

Nominal Output Rating of Generator In Watts	Average Monthly Wind Speed in MPH					
	6	8	10	12	14	16
50	1.5	3	5	7	9	10
100	3	5	8	11	13	15
250	6	12	18	24	29	32
500	12	24	35	46	55	62
1,000	22	45	65	86	104	120
2,000	40	80	120	160	200	235
4,000	75	150	230	310	390	460
6,000	115	230	350	470	590	710
8,000	150	300	450	600	750	900
10,000	185	370	550	730	910	1090
12,000	215	430	650	870	1090	1310

Modern wind generating plants use two or three long blades resembling an aircraft propeller. These propellers operate at a high tip speed relative to the wind velocity. Typical tip speeds run between 6 to 8 for high efficiency propellers. Slower running multiblade water-pump windmills have a speed ratio from 1 to 3. While slower running, multi-blade windmills have a higher starting torque and are smoother at low wind speeds.

Because conditions such as temperature and turbulence vary greatly, many wind generator manufactures will not commit to specific outputs. However to get a rough estimate of the feasibility of a wind powered installation the preceding table has been assembled. It should not be considered as an absolute certainty but instead as what *might* be possible.

BIO GAS – SEWAGE GAS – DIGESTER GAS: Gas produced by digestion, known as marsh gas, land fill gas, sewage gas or bio-gas is approximately 54 to 70% methane and 29% to 45% carbon dioxide. The fuel value of bio-gas is directly related to the amount of methane it contains. Methane has a heating value of 910 Btu / ft^3. The composition of the gas depends upon the raw materials with plant matter forming a higher percentage of CO_2, making a lower heating value gas.

The gas producing reaction in the digester is $2C + 2H_2O = CH_4 + CO_2$. Digesters or sewage plants will produce .8 to 1 ft^3 of gas per day per person served by the plant. Chickens produce enough manure to produce about 2 ft^3 of gas per day per chicken. A large plant could feasibly produce enough gas to run itself. Bio-gas may supplement natural gas and reduce the cost of operation. However you would be doing much shoveling if you wanted to power your house solely on bio-gas.

A "rotten-egg" smell indicates the presence of hydrogen sulfide in the fuel gas. Sulfur oxides, resulting from combustion, will combine with water to form sulfuric acid. The water dew-point is typically between 100° to 150° F and the acid dew-point is between 200° to 300° F. To prevent condensation and corrosion of the exhaust system, the exhaust gas is kept above 350°F and the inlet water on exhaust gas heat exchangers is usually 160° or above. Heat exchangers and exhaust piping may be built from stainless steel or the fuel gas may be filtered to remove the sulfur.

Commercial landfill gas operations typically chill the gas 40° F to condense the water vapor and then carbon filter it to remove the siloxanes. The silicon-based siloxanes come from the decomposition of cosmetics, hairsprays, soaps, creams and dry-cleaning products. Siloxanes will eventually coat the combustion chambers of the engines if not removed from the gas supply

4. COMPARISON OF GENSET TYPES:

SPARK IGNITION SETS: This category includes both gasoline and gas fired sets. While there are high-compression gas sets with high efficiencies, they are uncommon in smaller sets. This does not mean that you can not build one, however it would involve making a new set of pistons, which is described in another book.

Gasoline Engine Advantages: Small, light, fuel readily available, inexpensive, easy to convert to propane operation. Generally air cooled types available.

Disadvantages: Fuel quickly goes bad. Carburetor gums up if not operated weekly. High cost of fuel. Inefficient. High cost of operation. Exhaust fumes objectionable, very noisy. Short life-often a few hundred hours, considering that there are 24 hours in a day, that is not very long in a situation where power is out for a few weeks. Few models run at 1800 rpm. Poor part load efficiency.

Propane Advantages: Low to moderate cost, fuel stores indefinitely, 3600 and 1800 rpm sets available, air and water cooled types available. Clean burning-exhaust is not objectionable, may be very quiet, some have long life. Lower initial cost if gas is already installed for other appliances. Good to excellent choice for back up power.

Disadvantages: Low efficiency unless engine is specifically designed for high compression. High cost of operation. High cost of large tanks. Plan on spending $1000 to $1500 for a large tank and $3 per gallon to fill it. Poor part load efficiency.

DIESEL SETS:

Advantages: Very efficient, lowest cost of operation. Good part load efficiency. Burns a wide variety of fuels. Best for long term operation. Heavy construction. Fuel stores well. Fuel not explosive. Low fire hazard. Most are 1800 rpm.

Disadvantages: Very noisy, requires noise abatement. High vibration. High initial cost. (however, not much difference if considering the cost of a large propane tank) Fuel and exhaust smells objectionable. Heavy, not portable.

5. SIZING YOUR GENERATOR:

Generators may be sized to run your whole house or only the essential loads. The bare essential loads include the refrigerator, well pump, minimal lighting and perhaps a hotplate.

Typically, the largest load or motor to be started dictates the minimum size of the genset. Out of the bare essentials list, the refrigerator or well pump would be the critical load. In Florida, the air conditioner is usually the largest household load and may be considered essential in the sweltering summer heat, however small window units can serve in an emergency and allow the use of a smaller generator set. Air-conditioners and motors are described in separate chapters. The water heater and electric range are the next largest loads. Refrigerators require approximately 2700 watts to start and 900 watts to run. Larger units may require more power. Lighting loads are usually small, at most 1 to 1.5 kW, which equals 10 to 15 100-watt lamps. Use of energy saver and fluorescent lamps may cut the lighting load by more than half.

Management of the electrical loads allows a smaller generator to be used. This means don't try to run your water heater and electric range at the same time, or turn off your AC unit when using other large electrical loads. Often, essential loads are moved to a separate breaker box that can be switched between utility power and the genset. The switching device is called a transfer switch that may be a manual switch or automatically start the generator and switch the loads.

If you are running your whole house on the generator, then the genset should be sized to the starting watts of the largest load plus the usual expected load. A summary of typical appliance electrical requirements is found on the next page.

18

Typical Electrical Loads: (AMPS X VOLTS = WATTS)		
Appliance	Starting Watts	Running Watts
Coffee Maker	600	600
Dishwasher-cool dry	750	250
Electric Fry Pan	1500	1500
Stove - Large element	2400	2400
Small element	1200	1200
Micro wave oven Standard Size	1500	1500
Microwave Oven Small- 650 watts	1000	1000
Refrigerator or Freezer	2700	900
Automatic Washer	1200	1200
Clothes Dryer	7200	6000
Electric Water Heater	6600	6600
Incandescent Lamp	as indicated on bulb	as indicated on bulb
Radio	50 to 200	50 to 200
Television	350	350
Garage Door	350	350
Well Pump, 1hp	5700	1900
Desktop Computer	250-350	250-350
CRT Monitor	100	100
LCD Monitor	40-50	40-50
Laser Printer	400-600	400-600
Ink-jet Printer	75	75
Laptop Computer	50-75	50-75

6. MOTOR LOADS AND MOTOR STARTING:

Air conditioners, refrigerators, fans, pumps, compressors and elevators all have motors, making motor starting and running one of the main applications of a generator set.

Motors draw large currents when started at full voltage. A typical motor draws about six times its nameplate "Full Load Current" when starting and continues to do so until it reaches about 75% of its rated speed. This usually lasts only for a fraction of a second to a few seconds. The current required to start a motor at full voltage is close to the "Locked Rotor Current" which is usually listed on the motor nameplate. NEMA, The National

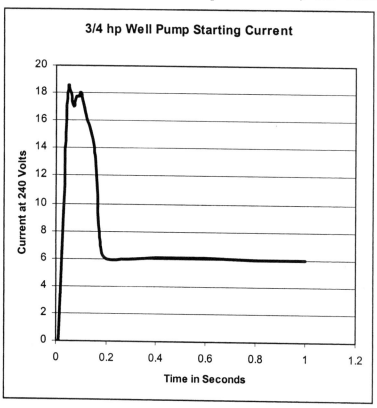

Electrical Manufacturer's Association, uses a code letter to classify motors by the locked rotor current per horsepower.

Small Motor Locked Rotor Codes		
Size HP	Code	Locked Rotor kVA / Hp
1 to 2	L or M	9 to 11
3	K	8 to 9
5	J	7 to 8
7.5 to 10	H	6 to 7
15	G	5.6 to 6.3

While the theoretical current required to start a motor is quite large at full voltage, reducing the voltage can considerably reduce the starting current. In real life, there is always some voltage dip in the line when starting a motor. Even when using utility power, as my home air-conditioner starts, the lights in my house briefly dim, indicating a voltage dip. When the voltage drops, current is also reduced so that the starting kVA (volts x amps) are reduced by the square of the voltage drop. A 30% voltage drop reduces the kVA by about 50%, .7 volts x .7 amps = .49kVA.

Motors are controlled by relays or contactors. Voltage dip affects the motor control relay's ability to stay closed and it may begin to chatter if the voltage dips too much. Most motor contactors (relays) will stand a 40% voltage drop. However some relays will chatter at 20% voltage drop. The drop out voltage of control relays may be found from the manufacturer. Adding a rectifier and capacitor in line with the relay, (running it on DC) may prevent the relay from chattering. Generators may be sized for a particular load by the "acceptable voltage drop."

A general rule of thumb is that the generator set should be sized to 3 times the running current of the largest motor plus the expected base load or usual load.

Motor Full-Load Current, Amps					
Single Phase			**Three Phase**		
hp	115 v	230 v	208 v	230-v	460 v
0.17	4.4	2.2			
0.25	5.8	2.9			
0.33	7.2	3.6			
0.50	9.8	4.9	2.2	2	1
0.75	13.8	6.9	3.1	2.8	1.4
1.00	16	8	4	3.6	1.8
1.50	20	10	5.7	5.2	2.6
2.00	24	12	7.5	6.8	3.4
3.00	34	17	10.6	9.6	4.8
5.00	56	28	16.7	15.2	7.6
7.50	80	50	31	22	11
10.00	100	50	31	28	14

7. AIR CONDITIONING AND REFRIGERATION:

Rating of Air conditioners:

Air conditioners are rated in Btu per hour for the smaller sizes and by tons in central heating and air systems. Although the systems of measurement appear different, they both related to Btu.

One ton of refrigeration is equivalent to the amount of heat required to melt 1 ton (2000 pounds) of ice to water at 32° F. It takes 144 Btu per pound to melt ice into water. Therefore 1 ton of air-conditioning is equivalent to: 144 Btu x 2000 pounds = 288,000 Btu. per 24 hours. Refrigeration calculations are based on a 24 hour day.

1 ton of air conditioning = 288,000 Btu per day

= 12,000 Btu per hour

= 200 Btu per minute

Because water must be cooled from room temperature, frozen and cooled further, the ice making capacity of a refrigeration plant is only about 60% of the ton rating. Therefore, a 1 ton refrigeration plant will make: 2000 lbs. x .60 = 1200 pounds of ice in 24 hours.

ELECTRICAL REQUIREMENTS OF AIR CONDITIONERS:

Air conditioners are likely to be the largest single load on a small power generation system and therefore may determine the minimum size of the generator set. While the power required to run an air conditioner is not particularly large, the current required to start the compressor is quite large. See the section regarding motor starting for more details. For emergency power systems, the generator capacity required to start central air conditioners may make the system too large and expensive. Substituting a few smaller window units will allow the use of a smaller generator set without too much loss of comfort. A summary of typical air conditioner capacities and starting currents are listed below.

Capacity and Electrical Requirements of Air Conditioners						
Tons	Btu / hour	Run Amps	Run Watts @220 volts	Locked Rotor Current amps	Starting Watts	3x running watts
1	12,000	7.3	1,600	38	8,500	5000
2	24,000	10.3	2266	56	12,000	6800
2.5	30,000	13.5	2970	72.5	15,950	9000
3	36,000	15.4	3388	88	16,500	10,200
3.5	42,000	18	3960	104	21,000	12,000
4	48,000	23.7	5214	129	26,000	16,000
5	60,000	28.9	6358	169	31,800	19,000
Electrical data is approximate. Starting watts based on 5x running current.						
Unit might start with as little as 3x running current.						

8. TRANSFER SWITCH:

A transfer switch allows the power input to be switched from utility power to the generator set. A transfer switch may be small, switching only a few loads or it may switch the entire house. An automatic switch starts the generator and switches the loads when the generator comes up to speed during a power failure and then turns off the generator when the power is restored. Manual switches are operated by hand. At no time are both power sources connected to the loads at the same time. Transfer switches must be used to prevent generator power from back feeding into the power lines where electrical workers may be electrocuted. It is both dangerous and a violation of the electrical code to apply a generator to a household load without being disconnected from the utility power. Likewise, the transfer switch prevents utility power from going through your generator.

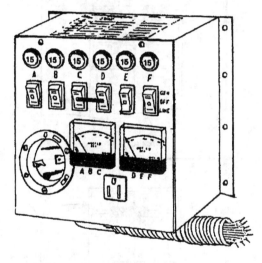

Whole house transfer switches are installed before the main breaker while the part load switches are installed after the main breaker in a household system.

Small switches for part load transfer may be purchased or assembled. A commercial part load manual transfer switch is seen above right. Part load switches are usually rated from 30 to 60 amps depending upon the number of circuits served. A small, part load transfer switch may be assembled from a breaker box by tying the two power input breakers together so that only one breaker may be turned on a time.

Whole house transfer switches are usually rated 200 amps. They may be a separate switch or included in the meter box or main electrical panel. When a separate switch is installed before the main breaker, the feeder cable from the switch to the main breaker box must be upgraded from 2 conductors with a metallic wrap to 3 conductors with a wrap or ground. I am using a manual 200-amp Cutler-Hammer transfer switch available from Harbor Freight Tools. It is a large switch measuring 38 x 19.5 x 6-inches.

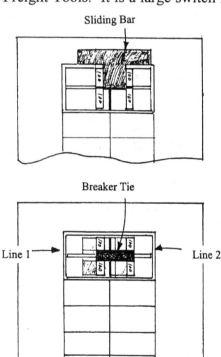

Sliding Bar

Breaker Tie

Line 1

Line 2

You will need all of the room inside the box when working with large feeder cable.

Left: Various methods of tying two breakers together to make a small transfer switch.

26

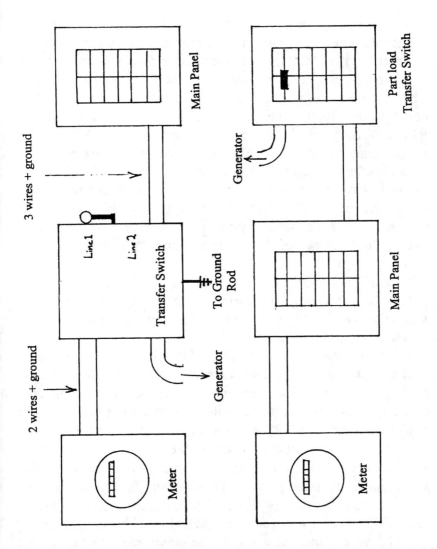

Two types of transfer switches. The top drawing switches the entire electrical panel between two power sources. The bottom drawing is for transfer of essential loads only.

27

BACK-FEEDING A PANEL IN EMERGENCY SITUATIONS:

There may be times when a generator is available and there is no transfer switch. Multiple extension cords pose a hazard or there may not be enough extension cords to properly supply the loads. In such cases the panel may be back fed through an existing circuit. The most important issue here is to TURN OFF THE MAIN BREAKER before the generator is connected and LEAVE IT OFF until the generator has been fully disconnected from the circuit.

The breaker panel may be powered through an unused double pole 120/240-volt circuit breaker or by using a double male plug and connecting to the dryer circuit. Typically, #6 wire is the smallest for use in back-feeding a dryer plug. The generator capacity must not exceed the capacity of the dryer outlet, which is usually 60 amps.

Make all connections BEFORE starting your generator. You do not want to be walking through the dark with exposed live cables. *Turn off the circuit. Make the connection. Start the generator, then turn the circuit breaker on.*

If you are working with a small 120-volt generator, the main panel may still be powered, however you will not be able to power any 240-volt circuits. Again the main breaker and any double pole breakers must be turned off. Double male extension cords may be plugged into an open 120-volt socket. It will require 2 separate cords into 2 separate sockets to power all of the 120-volt circuits. It is critical that the plugs are properly wired on both ends and that they maintain the same polarity. These should be checked with an ohmmeter to be absolutely certain. The extension cords should be made of 12-gage wire. No cord should be loaded beyond 20 amps.

Under no circumstances should the main breaker be turned on while that generator set is connected by any of the above mentioned methods.

If you don't fully understand what you are doing, then don't do it.

28

9. OPERATON of MULTIPLE GENERATORS:

PARALLEL OPERATION OF GENERATORS:

In most commercial power plants, power is supplied by several small units rather than one large unit, this is called parallel operation. Generators may be run in parallel for a number of reasons. 1. Several smaller units are more reliable that one large unit. 2. The units may be connected or taken out of service as the load on the plant varies. This keeps the generators loaded up to their rated capacity, where they are most efficient. 3. A unit may be taken out of service and repaired without the total loss of power. 4. The load may exceed the rated capacity of any single available unit.

Generally, for small emergency systems, the generator runs a few essential loads leaving the remaining circuits with no power. However, one may have multiple small generators and consider their 5-ton air conditioner an essential load. While no single small generator will start such a large air conditioner, it may be possible to parallel multiple small generators and start the load.

A brief discussion of the principles involved is required before connection of the machines.

1. The voltage, frequency and polarity of the connected machines must be the same. Looking at the label is not enough. The generators must be checked with a meter. One generator running at 63 Hz and 240 volts and one with 57 Hz and 210 volts are *not* the same and adjustments to the engine governors and perhaps the voltage regulators may be required.

2. The engine governors must be "droop" type control where the engine speed decreases with the load. Mechanical governors are droop type. Most small generator sets have this type of control. Some gensets may have electronic governors were speed droop is a selectable option. If all the generator sets in a system have droop governors, then the generators can be set to share the load equally.

3. Connection must be made at the instant where the voltage, frequency and phase of the connected units are the same. Once connected, the units will stay in phase. If two generators (alternators) connected in parallel attempt to pull out of step, a current is developed that circulates between the machines. This current speeds up the slower or lagging machine and slows the faster or leading machine thereby acting to prevent the machines from pulling out of synchronism. This is called the synchronizing current. If the alternators are operating under a load, the leading alternator takes more of the load and the lagging alternator takes less.

SETTING UP SETS FOR PARALLEL CONNECTION:

Assuming that you are connecting smaller gen-sets together the following sequence is used:

1. Using a load bank, the first generator is heavily loaded and the engine speed is adjusted to 60 Hz. (Check for over-speed at light loads and adjust the governor for an acceptable compromise. Usually ± 3 Hz is acceptable.) While you could purchase a load bank, an old electric stove makes a good load. Old stoves are routinely discarded and usually free for the taking. I have two for testing my gen-sets.

2. Adjust the speed of the oncoming gen-set to match the speed of the first set.

3. Synchronize the second set and close the breaker. Lamps may be used to determine the difference in phase or frequency if connected across the poles of a breaker. The lamps will flicker at a frequency equal to the difference in frequency between the two generators. As the frequency of the second set approaches the first set, the flicker will become slower and slower. When the lamps go dark, the breaker may be closed. While a voltage difference may still exist between the two sets, the dark lamps indicate that it is fairly

low. When connecting across a 240-volt line, two 120-volt lamps are connected in series.

Three-phase synchronizing is similar. If the 3 lamps grow bright and dim together, then the alternator is properly connected. If they dim and brighten in sequence, then the phase rotation is opposite and any two leads must be reversed.

4. Using an amp-meter, increase the speed of the incoming set to assume half of the existing load.

5. Readjust the speed of both sets to achieve the proper loaded frequency. Here you can add the second stove to check for proper frequency and load sharing.

Once the governors are set, future parallel connections may be made with little or no adjustment, closing the breaker only when the lamps are dark.

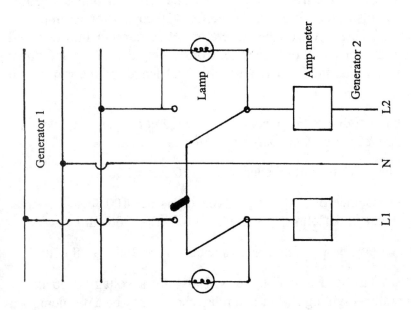

10. VOLTAGE DROP IN FEEDER LINES:

Most wiring books have a wire table that specifies the size of conductors for a particular amperage. These tables work well for short runs but the resistance of the feeder cables should be considered if the generator set is located more than a very short distance away from the electrical panel. There is a voltage drop and a power loss when using long feeder cables. At lower voltages, both voltage drop and power loss can be considerable. Battery cable losses dramatically affect inverter performance.

Depending upon the location of your generator set, the length of your feeder cables can quickly add up to a few hundred feet. Ignoring the resistance of the feeder cables may give you disappointing performance at full load or when starting motors.

Example: Assume that a generator set is located approximately 50 feet from the back of a house. The electrical panel is located on the garage wall at the front of the house.. The gen-set supplies 30 amps at 240 volts. The "wire table" specifies #10 copper for 30 amps. The straightest run to the generator set is through the attic and down the back wall of the house. Considering the run up and down the walls plus the buried cable, the total length of the run is 150 feet.

The net voltage at the panel is the voltage at the generator less the voltage loss in both the outgoing and return wires.

The length of the both feeders is 2 x 150 feet = 300 feet

The resistance of #10 wire is .102 ohms per 100 feet. The total resistance is .102ohms x (300feet / 100 feet) = .306 ohm.

The voltage drop = amps x resistance = 30 x .306 = 9.18 volts.

When starting a motor, there will also be a voltage drop in the generator. Lighting flicker and relay chatter may be a problem.

Annealed Solid Copper Wire

Gage Number	Diameter, mils	Area		Ohms per 1000 Feet		Ohms per Mile	Pounds per 1000 Feet	Pounds per Mile
		Circular Mils	Square inches	77 F	150 F			
0000	460.0	212,000	0.166	0.05	0.0577	0.264	641	3384.5
000	410.0	168,000	0.132	0.063	0.0727	0.333	508	2682.2
00	365.0	133,000	0.105	0.0795	0.0917	0.42	403	2127.8
0	325.0	106,000	0.0829	0.1	0.116	0.528	319.9	1689.1
1	289.0	83,700	0.0657	0.126	0.146	0.665	253	1335.8
2	258.0	66,400	0.0521	0.159	0.184	0.839	201	1061.3
3	229.0	52,600	0.0413	0.201	0.232	1.061	159	839.5
4	204.0	41,700	0.0328	0.253	0.292	1.335	126	665.3
5	182.0	33,100	0.0260	0.319	0.369	1.685	100	528.0
6	162.0	26,300	0.0206	0.403	0.465	2.13	79.5	419.8
7	144.0	20,800	0.0164	0.508	0.586	2.68	63	332.6
8	158.0	16,500	0.0130	0.641	0.739	3.38	50	264.0
9	114.0	13,100	0.0103	0.808	0.932	4.27	39.6	209.1
10	102.0	10,400	0.00815	1.02	1.18	5.38	31.4	165.8

Annealed Solid Copper Wire

Gage Number	Diameter, mils	Area		Ohms per 1000 Feet		Ohms per Mile	Pounds per 1000 Feet	Pounds per Mile
		Circular Mils	Square inches	77 F	150 F			
11	91.0	8,230	0.00647	1.28	1.48	6.75	24.9	131.5
12	81.0	6,530	0.00513	1.62	1.87	8.55	19.8	104.5
13	72.0	5,180	0.00407	2.04	2.36	10.77	15.7	82.9
14	64.0	4,110	0.00323	2.58	2.97	13.62	12.4	65.5
15	57.0	3,260	0.00256	3.25	3.75	17.16	9.86	52.1
16	51.0	2,580	0.00203	4.09	4.73	21.6	7.82	41.3
17	45.0	2,050	0.00161	5.16	5.96	27.2	6.2	32.7
18	40.0	1,620	0.00128	6.51	7.51	34.4	4.92	26.0
19	36.0	1,290	0.00101	8.21	9.48	43.3	3.9	20.6
20	32.0	1,020	0.000802	10.4	11.9	54.9	3.09	16.3
21	28.5	810	0.000636	13.1	15.1	69.1	2.45	12.9
22	25.3	642	0.000505	16.5	19.00	87.1	1.94	10.2
23	22.6	509	0.000400	20.8	24.00	109.8	1.54	8.1
24	20.1	404	0.000317	26.2	30.20	138.3	1.22	6.4
25	17.9	320	0.000252	33.0	38.10	174.1	0.97	5.1
26	15.9	254	0.000200	41.6	48.00	220	0.769	4.1

11. MOUNTING YOUR GENSET:

There are several factors to consider when selecting a site for your generator set:

1. Noise
2. Protection from the elements and vandals
3. Accessibility for maintenance
4. Fuel storage
5. Cost of feeder cable
6. Odors from fuel and exhaust

NOISE: Without noise abatement, exhaust and mechanical noise varies from 90 to 110dBA at 6 feet from the engine. The cooling fan is a significant source of noise. There are several grades of mufflers available, which easily control exhaust noise. Industrial mufflers reduce the noise somewhat. Residential grade mufflers limit the noise to 95 dBA and critical mufflers limit the noise to 85 dBA. Point the exhaust away from the house and neighbors. Intake noise is dramatically reduced with a muffler.

Noise drops quickly with distance, doubling the distance reduces the noise by about 5.5 dB (Note: because a generator is not a point source, this rule starts at about 30 feet). About 65 – 70 dB is considered a normal conversation.

PROTECTION FROM THE ELEMENTS: The elements include storms, snow, heat, sand, floods and vandals or burglars. The genset should be high enough not to be flooded. FEMA maps show the probability and height of floods. If the weather is cold, the direction of the prevailing wind should be considered. Locate the genset so that snow does not pile up against the access doors or blow into the shed when the doors are open. Regarding storms, the question here is does the enclosure need to be weather resistant, keep the rain and snow off or weather proof, able to withstand 115+ mph hurricane winds?

ACCESSIBILITY FOR MAINTENANCE AND SERVICE: Maintain a minimum clearance of 3 feet on all sides of the unit. Removal for repair or replacement should be considered when selecting a site. You must consider how the equipment will initially be set in place. If located in a building, a hatch in the roof large enough to move the genset through might be an option.

FUEL STORAGE: Type and location of the fuel storage is important. If you are storing a large quantity of fuel, more than you want to carry in a can, accessibility of the tank is important. Most fuel trucks carry a 50-foot hose, although some carry only a 20-foot hose. Be sure that the driver can get close enough to your tank. Some waste oil trucks are gravity drained, meaning that their tank needs to be a little higher than your tank.

Larger tanks may be located for accessibility and smaller "day tanks" may be located close to the set. The day tank is filled by pumping from the larger tank.

FEEDER CABLE: The feeder cable should be sized to carry 100% of the load. If there is no breaker located on the set, then it should be sized to 115% of the maximum output. While the cost of feeder may be significant, it should not be the only consideration in the location of a set. Over time, the cost of feeder may seem trivial relative to the annoyance of a poorly located set.

ODORS FROM FUEL AND EXHAUST: Exhaust gas is poisonous and must be pointed away from people and animals. Do not let exhaust gas blow into the eves of a building. Diesel exhaust is usually objectionable, however propane fueled sets do not produce an odorous exhaust. Fuel smells may also be objectionable. Locate the set where no one will be bothered by fuel smells.

ENCLOSURES: The best enclosure is a concrete block or brick building. You can often purchase pallets of odd lot or leftover bricks at a brickyard for pennies each, provided you agree that the sale is final and there are no returns.

Sheet metal enclosures, while used in all climates, are more suitable for use on emergency gensets in moderate climates where temperatures are above 30° F and not subjected to dust or insects. Salt spray is particularly hard on metal enclosures. Stainless or painted aluminum are recommended for salt resistance.

If weather resistance is all that is required, wooden enclosures may be used. Vibration isolation is especially important in this situation. Wooden walls tend to vibrate, broadcasting the noise. If the shed has a wooden floor, it must be isolated from the walls. The floor containing the genset must not touch the foundation of the walls or the rest of the building at any point. Offset studs may be used to reduce the transmission of noise to the outside walls.

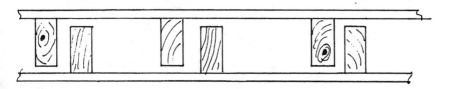

Offset Studs

MOUNTING PADS: The best surface for mounting a genset is a concrete pad. The mounting pad should extend six inches beyond the genset base dimensions. The pad is poured over a 4-inch deep bed of rock. This rock bed should surround the pad by 8 to 10 inches. The rock increases the vibration isolation so that it is not transmitted through the ground to other structures. It also provides drainage.

Generally, pads are poured from 2500 to 3000 psi concrete. Redimix bag concrete and a small mixer work well for this type of job. The concrete is reinforced with eight-gage wire cloth or fiberglass. The minimum weight of the pad should be at least equal to the weight of the genset. 150% would be better. Concrete weighs about 144 lbs. / ft³. A 4-inch slab weighs about 48 pounds per square foot.

MAKING A GENERATOR BASE:

The generator base is made from a 20-foot length of 4-inch C-channel iron. It may be configured as a simple base or enclosed for use as a fuel tank. It may also have a mesh of welded rebar and be filled with concrete for use as an inertia block in critical vibration applications. The crosspieces are positioned under the engine and generator mounts as required. The radiator is mounted on a welded "U" shaped section of 2-inch by 1/8-inch angle iron. A thin sheet metal plate is added to the bottom of the radiator frame to reduce the recirculation of cooling air. The assembly bolts to the base.

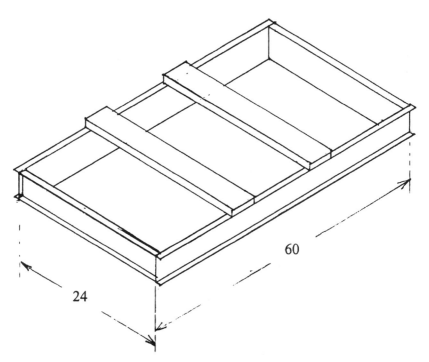

60

24

The enclosed generator base holds approximately 24 gallons of fuel. The channel is notched to position a drain plug as low as possible, a drain hole is drilled and a pipe coupling is welded in place. 16ga sheet is used to cover the top and bottom. The filler neck is made from a pipe nipple. Commercial truck caps are available and are recommended. They are made of bronze and are non-sparking. They also have fusible plugs that melt out providing a safety vent. I have also used a pipe cap, drilling a few pinholes in it and greasing the threads.

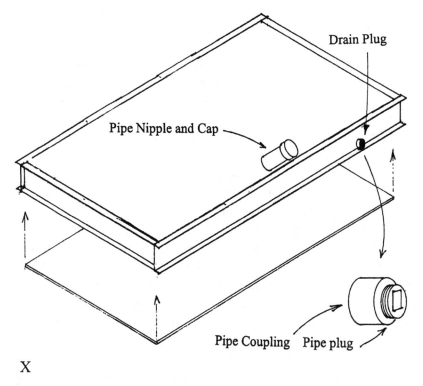

Drain Plug

Pipe Nipple and Cap

Pipe Coupling   Pipe plug

X

Note: I added a few fork cutouts to my base (non-fuel tank type) to allow my forklift to pick it up. While they do work, they are probably a waste of time in that they are small and difficult to properly stab with the forks. I bolted two 4 x 4's to the bottom and it is much easier to lift. The torch cut fork holes caused the C channel to warp. I do not recommend cutting fork holes in the channel.

BUILDING METAL ENCLOSURES: A sturdy enclosure may be built from angle iron and sheet metal. Having access to a metal shear and break will make this little more than a one-day job. I paid a visit to the local junior college where I was able to use the equipment. If you do not have access to such equipment, the job takes a little longer, with most of the time is spent cutting the sheet metal.

The enclosure frame is made from 1/8-inch by 2-inch angle and the sheet metal is 16-gage steel. This makes a very sturdy enclosure that has weathered several hurricanes. The front of the enclosure has a hinged door with a section of expanded metal over it. Because this is an emergency set. I did not attempt any noise abatement. While the generator is not particularly noisy, the fan noise is substantial and sounds similar to our air-conditioning unit. If I were operating a radiator type set on a full time basis, I would certainly do something about the fan noise.

The set has been out there for 8 to 10 years and only problem that I have had is the top panel rusting. If I had it to do over again, I would try aluminum or perhaps stainless steel. Because commercial restaurant equipment is made from stainless, a local restaurant supply may be able to direct you to a stainless sheet supplier. Because all you want is rust protection, 400 series metal is fine. Considering all the time over the years I have spent sand blasting patching and painting the top, this is probably money well spent. Another change that I would make to the original enclosure would be to angle the top slightly so that rainwater drains off. This modification alone may prevent the rust problem.

The genset is mounted on a 4-inch concrete slab. It is raised up 2 blocks high and these blocks are filled with concrete. It is high enough that it is easy to work on and rainwater does not splash up into it. I am very pleased with the height and would recommend elevating your set. Rubber waffle pads sit between the genset frame and the blocks. Electrical feeder was run through 2-inch plastic conduit, buried and brought up through the base before pouring.

The genset was loaded into the back of a pickup where it was pulled up as close as possible to the base. An engine hoist and leveler were used to lift the set and pull it into position. J-bolts or concrete anchors were put through the mounting holes in the base of the genset. The concrete bocks were then filled with wet concrete and the genset lowered into position, pushing the J bolts down into the wet concrete. The surface was smoothed and left to harden. By setting the bolts this way, I knew that they would fit.

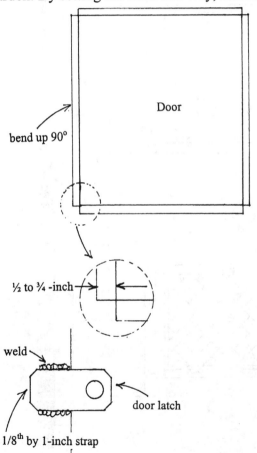

Left: The door detail. The ends of the sheet metal are notched for the bend. Short strips of 1/8th-inch strap are welded in place as door latches. Bolt holes are drilled in the latches and nuts are welded inside the frame.

Small hinges are welded to the door. These are zinc plated so the zinc is ground off the around the bolt-holes and spot welds are made in these holes.

41

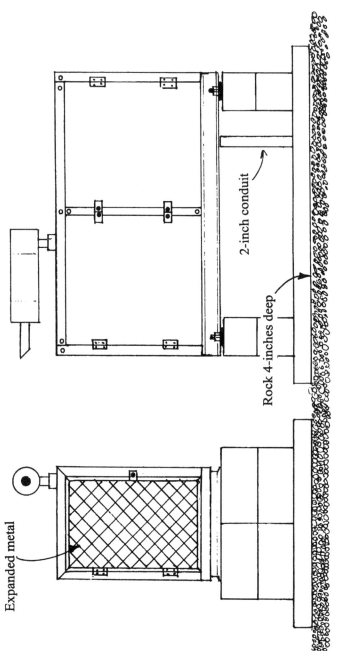

Expanded metal

2-inch conduit

Rock 4-inches deep

Note: Blocks are reinforced with rebar driven into the ground before pouring 4" slab.

42

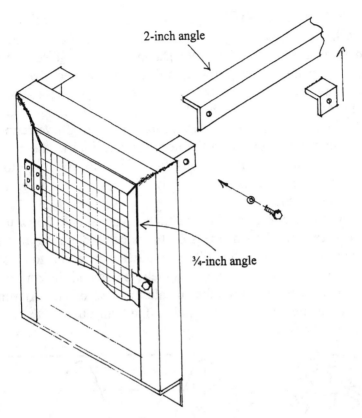

2-inch angle

¾-inch angle

End Frame and Door Detail

Exhaust pipe sleeve prevents rain on the top panel from draining into the enclosure. Tabs are alternately bent so that they hold the sleeve tightly in the panel. The assembly is then caulked. Slope top panel away from the exhaust 1/4 –inch per foot.

12. BASIC ELECTRICITY: Two types of electrical circuits concern the generator project. Direct Current or DC circuits and Alternating Current or AC circuits. Batteries produce direct current and the power from your wall socket is alternating current.

DC may be produced from DC generators, batteries, and rectified AC (a rectifier converts AC to DC). Current in a DC circuit flows from the positive terminal to the negative terminal, however electrons flow from the negative terminal to the positive terminal. Within the battery, ions flow from the positive to the negative electrode. This ion flow is of importance in galvanic corrosion problems.

AC flows in one direction and then reverses and flows in the opposite direction. This happens 60 times per second in your wall outlet and therefore is called 60-cycle power, or is said to be at 60 Hertz (Hz). Both the number of magnetic poles and the rpm of the alternator determine the number of alternations or Hz. Alternator poles are magnets and come in pairs, one being the north and the other being the south pole.

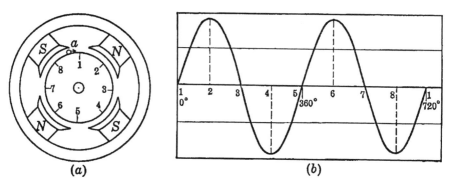

(a)   (b)

As the wire *a* above moves under the magnets from positions 1 to 8, a rising and falling voltage is produced as seen graphically on the right. Every time the wire passes under a pair of poles, one cycle is completed. A four-pole generator therefore completes two electrical cycles per mechanical revolution. One cycle being 360 *electrical degrees.* The frequency in cycles per second equals the number of *pairs* of poles passed in one second.

Frequency (F) = (P x RPM) / 120   where P = number of poles

RPM = 120 F / P   P = 120F / RPM

Example: Find the frequency of a 2 pole 3600 RPM generator and find the rpm of a 4 pole, 60 Hz generator:

F = (2 x 3600) / 120 = *60 Hz*, RPM = (120 x 60) / 4 = *1800 RPM*

MODIFIED SINE WAVES: The voltage in a sine wave moves smoothly from zero volts to the maximum and back down zero. A "modified sine wave" comes on at full voltage. Stays there for a moment and then cuts off, dropping straight to zero, where it stays for a moment. A square wave comes on at full voltage and then switches to the opposite polarity with no zero point in between.

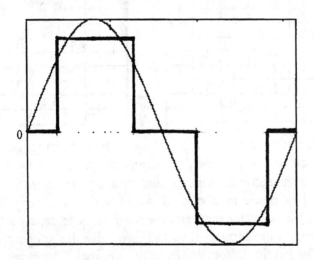

Sine and Modified Sine Waves

Many inverters, which are used to convert DC from batteries into AC, generate a modified sine wave. While the modified sine wave works well in resistive and most transformer circuits, it does not work very well in any circuit that requires voltage control such as a dimmer or speed controlled motor. Many computers use "switching power supplies" which require voltage control and may not work on the modified wave form.

ALTERNATING CURRENT AMPERE: Seen below left is an alternating current sine wave, having a maximum value of 1.414 amp. It might seem that the number value of amps should be based on the average of the value; however if the average of the complete cycle is taken, the average value is zero. There is just as much negative as positive current.

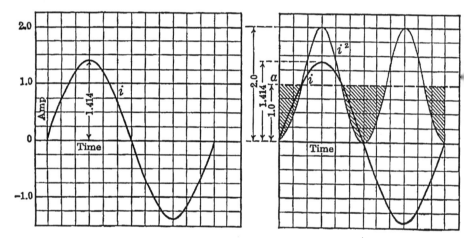

The alternating current amp is based on its heating value. An alternating current amp, when flowing through a given resistance will produce the same amount of heat as a direct current amp. If a heating coil is place in a cup of water and 1 DC amp is sent through it and the water temperature rises 10 degrees, an AC amp flowing through the same coil would also raise the temperature 10 degrees. The heating effect varies as the *square* of the current or:

$$i^2 R \quad \text{where i = current and R = resistance}$$

The plot on the right shows the sine wave and its *squared* values. The *average* of the *squared* current would be 1 amp. (If the tops of the wave were cut off, they would fit into the valleys in between.) To find the value of the current, the square root of the *averaged square* current is taken. This is called the RMS current or the root mean square. An alternating current amp has a maximum value of 1.414 and the RMS current is 1 / 1.414 or .707 x the maximum current. This will be of more importance when discussing inverters.

AC GENERATORS:

When a conductor passes a magnetic pole, a voltage is generated in that conductor. However, the generation of voltage in an armature depends only upon the relative motion of the conductor and magnetic field so that either the armature or the magnetic field may be rotating. Because it is easier to insulate and take power from a stationary coil, typically the field rotates and the armature is stationary.

The voltage in the armature is controlled by, among other things, the strength of the magnetic field. Direct current passes through the field coils generating the magnetic flux. By varying the voltage in the field coils, a stable generator output voltage is possible with widely varying loads.

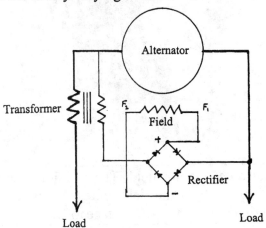

The field voltage is controlled by an automatic voltage regulator or a transformer on the generator output.

In a transformer regulated alternator, the output must pass through a transformer primary coil. As the load increases, a current proportional to the load is generated in the transformer secondary. In the most simple situation, this is fed back into the field coils to stabilize the alternator output. In another situation, this current is fed into another stationary field as part of a second alternator that supplies the field voltage. The second alternator is located on the main shaft behind the primary field. Its output is converted to DC and it supplies the voltage to the main field.

A brushless four-pole rotating field. The additional coils on the rear of the assembly supply the field voltage. A series of diodes mounted on the ring convert the AC voltage to DC.

Alternators may be the brush type or brushless. While the alternator output may be taken from a rotating armature through carbon brushes, brush type alternator usually means that voltage is supplied to the rotating field through carbon brushes. The photo above is of a brushless type of field.

AC generators for this project may be of the 2 or 4-pole type, the 2-pole type running at 3600 rpm and the 4 pole at 1800 rpm. To get the maximum power from an engine generator set, the 2-pole generator is selected. However, it is a much noisier set and not as efficient. This may not be an issue for short run emergency operation. After a hurricane, the power may be out for weeks; therefore I wanted a long running generator and selected the slow turning 4-pole set.

TEMPERATURE CLASSES OF INSULATION:

Generator insulation or varnish is classified by temperature and is separated into 4 groups, **A, B, F** and **H**. Temperature rating of insulation is given in degrees Celsius.

$$\text{Temperature } {}^{\circ}F = (1.8\ {}^{\circ}C) + 32$$

Insulation Temperature Limits at 40 Degrees C Ambient								
	Insulation Class							
	A		**B**		**F**		**H**	
	Cont	Stby	Cont	Stby	Cont	Stby	Cont	Stby
Maximum Temperature	100	125	120	145	145	170	165	190

DUTY CYCLES: There are two types of alternator duty cycles specified: *Continuous* and *Standby*.

49

Continuous Duty assumes operation for 24 hours per day, 7 days per week. The thermal life of the insulation is at least 20,000 hours.

Standby Duty allows a temperature rise of 25°C over the continuous duty rating. Operation at standby temperatures causes the insulation to age 4 to 8 times as quickly as the continuous duty rating. Therefore operation for 1 hour at standby rating equals as much as 8 hours at the continuous duty rating.

Overload Ability: A specification may require that an alternator operate a 10% overload for 2 hours in any 8-hour period. If the total output (110% rated capacity) is less than the standby rating of the alternator, then it meets the overload specification. Standby duty rated machines do not have an overload capacity.

Altitude Derating: Alternators are rated for duty up to 3300 feet. Above this altitude, derate the machine 1% for each additional 330 feet.

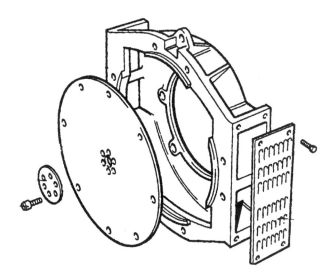

Generator Adapter and Steel Flex Plate

ALTERNATOR COUPLINGS:

Alternators may be of the single bearing type or two bearing type. The single bearing types have the front shaft bolted directly to the engine flywheel with a steel flex plate. Single bearing means that the generator head has only a rear bearing.

Two Bearing generators may be attached with a flex coupling, which is a steel ring and a keyed steel hub cast into a rubber ring.

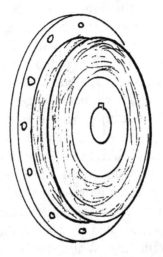

The Asian coupling depends upon rubber bushings over steel dowels that are inserted into holes in the flywheel.

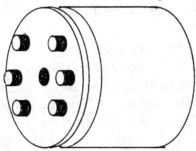

Coupling used on Asian 2 Bearing Generators

13. GENERATOR SUPPLIERS:

US suppliers include Marathon, LeRoy Somer and Newage. Of all the US suppliers, Newage proved to be both the lease expensive and the most reliable. Marathon, while providing a good quote, was out of stock for another six weeks. It required at least 5 phone calls over two weeks to LeRoy Somer to get a quote. They were the highest priced of all the suppliers. I also imported several Asian generator heads for evaluation. Transit time is 6 to 10 weeks. While they are much less expensive, the handling fees, once they are in the US, quickly absorb any savings. Because the fees are per port entry not per part, it would probably become cost effective if you were buying at least 6 to 10 generator heads at a time.

There are single and two bearing generator heads. Western (meaning US and European) generator heads usually conform to the SAE standards, while the Asian heads usually conform the China National standard. SAE heads are of the single bearing type.

While you can order an Asian generator head with an SAE coupling, they are not as common and much more expensive than the typical two bearing Asian generator heads. The standard two bearing head may be mounted with a flexible rubber coupling or the standard Asian coupling. While Asian couplings are very easy to assemble, any misalignment quickly wears the rubber bushings. Besides accurate alignment of the generator to the flywheel, polishing the mating holes in the flywheel should increase the life of the bushings.

Further comparison of the Asian and Western generators brings a few other notable differences. The finish of Asian parts is not as good as the Western parts however this is probably not an issue except at the mating surfaces of the generator adapter. These surfaces were covered with paint, which may or may not cause an alignment problem and early failure of the rubber bushings. I stripped the paint from this surface before assembly. The other notable difference is the thickness of the varnish on the coils. Heavy varnish secures the wire in the coils and prevents the wire from moving due to magnetic forces. It also prevents moisture from seeping between the sheets of the iron core causing rust and swelling. While the Western generators were very heavily varnished, the varnish on the Asian generator was much lighter. While I am sure that they were well insulated, I feel that the coils would loosen before those in their Western counterparts. I coated all of the coils with "Dolph's AC-46 Electrical Varnish" (about $5 per can). This may or may not improve the situation but it surely can't hurt. Another option would be to have the coils "dipped and baked" at an electrical shop. The prices vary considerable. A few calls yielded prices of $50 at a small shop to $350 at a large one for the same job.

I liked both the Asian and Western generator heads and they both do the job. Both types of heads easily start and run my 4-ton AC unit and have carried their full rated capacity for 8 hours into a resistive load consisting of several stoves. The Western parts were of higher quality but the Asian parts worked satisfactorily.

Exploded view and dimensioned drawings of the Stamford Newage Generator to follow. Note that the generator adapter may change. Consult the latest drawing revision and check your generator carefully.

Generator Selected for this Project:

Stamford Newage Model # BCI 164D1J 203A
120/240 volt dedicated single phase 4 pole 1800 RPM
16.5 kW capacity

Note: Generator quotes varied between 35% to 45% off retail price and included freight.

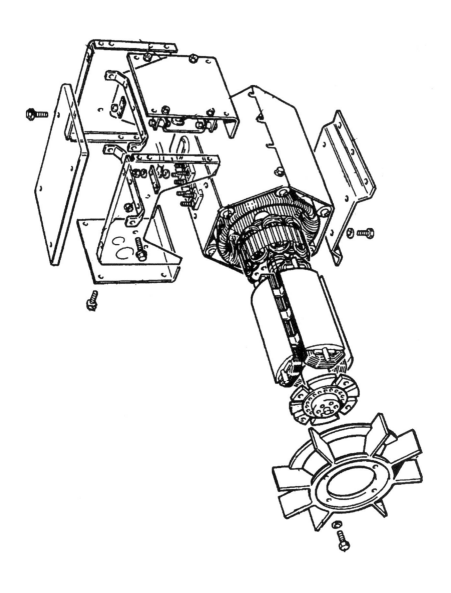

Exploded view of Stamford Newage 4-pole AC Generator

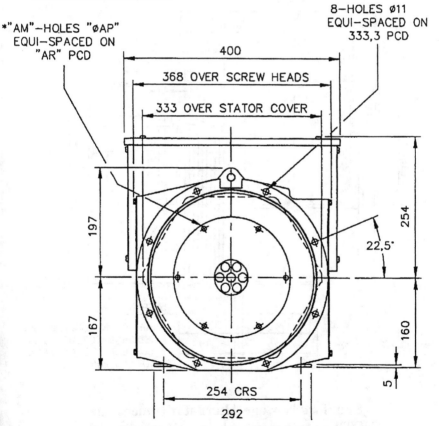

DIMENSIONS			
	FRAME	"A"	"B"
4-POLE	164 A	364,5	93
	164 B	364,5	93
	164 C	391,5	107
	164 D	391,5	107
2-POLE	162 D	364,5	93
	162 E	364,5	93
	162 F	416,5	132
	162 G	416,5	132

COUPLING DISC					
SAE	"AN"	"AM"	"AP"	"AR"	"V"
6,5	30,16	6	8,7	200,0	215,8
7,5	30,16	8	8,7	222,2	241,2
8	61,9	6	11	244,5	263,4

*"AM"-HOLES "øAP"
EQUI-SPACED ON
"AR" PCD

8-HOLES ø11
EQUI-SPACED ON
333,3 PCD

400

368 OVER SCREW HEADS

333 OVER STATOR COVER

197

167

254

22,5°

160

5

254 CRS

292

Stamford Newage Alternator Dimensions

	FRAME	NETT WT.(Kg)	C of G
4–POLE	164 A	79	182
	164 B	86	184
	164 C	94	196
	164 D	100	197
2–POLE	162 D	72	178
	162 E	82	181
	162 F	91	202
	162 G	105	203

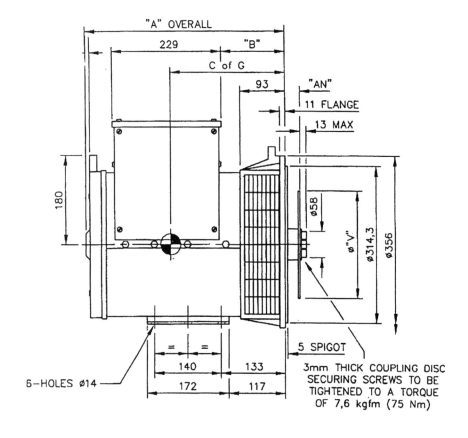

Stamford Newage Alternator Dimensions
NOTE: Some alternators have a 308 mm dimension
in place of the 314.3 mm dimension. NOTE: mm / 25.4 = inches

MARATHON part numbers Dedicated Single Phase, 4 - Lead, 60 Hz				
	kW Continuous Duty		kW Standby Duty	
Model Number	105 C	125 C	130 C	150 C
	Class F	Class H	Class F	Class H
281PSL1511	6.0	6.5	6.5	6.8
281PSL1512	7.7	8.2	8.2	8.5
281PSL1513	10	10.5	10.5	11
282PSL1514	12	13	13	13.5
282PSL1515	17	18	18	19
283PSL1516	20	22	22	23
283PSL1517	25	27	27	30
284PSL1518	30	32	32	35

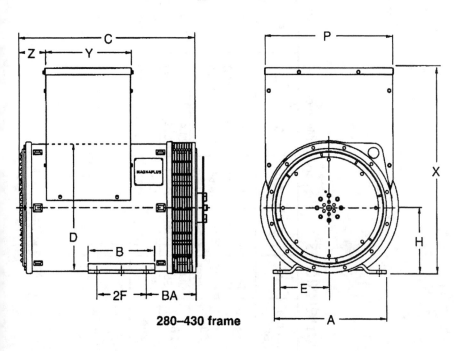

280–430 frame

Dimensions in inches and (millimeters)

Frame Size	A	B	BA	C	D	E	2F	H	P	X	Y	Z
281	14.00 (356)	7.00 (178)	6.56 (167)	15.95 (405)	13.75 (349)	6.25 (159)	5.00 (127)	8.00 (203)	13.42 (341)	18.56 (471)	6.86 (174)	3.50 (89)
282	14.00 (356)	7.00 (178)	6.56 (167)	17.94 (456)	13.75 (349)	6.25 (159)	5.00 (127)	8.00 (203)	13.42 (341)	18.56 (471)	6.86 (174)	3.50 (89)
283	14.00 (356)	7.00 (178)	6.56 (167)	20.44 (519)	13.75 (349)	6.25 (159)	5.00 (127)	8.00 (203)	13.42 (341)	18.56 (471)	6.86 (174)	3.50 (89)
284	14.00 (356)	7.00 (178)	6.56 (167)	22.44 (570)	13.75 (349)	6.25 (159)	5.00 (127)	8.00 (203)	13.42 (341)	18.56 (471)	6.86 (174)	3.50 (89)

14. GENERATOR PROJECT OVERVIEW:

The generator project may be divided into two types. The first type being a common radiator-genset used for both emergency and remote power generation. The second type of generator set recovers waste heat from both the engine water jackets and exhaust for use in an "off the grid" type power plant. While most users will be well served by the first type of plant, a few will find that heat recovery combined with the ability to burn waste oil products benefits them.

Converting a spark ignition engine to gaseous fuel operation involves little more than changing the carburetor and installing a gas regulator. More modern engines, which are computer controlled, may be converted as long as there is an engine driven distributor. With a small foundry you can cast a new distributor housing to mount a points type or HEI distributor on your engine. Making the coupling and bell housing to attach a generator to an auto engine, while a little time consuming, is neither difficult nor expensive. All of the parts that are cast for this project are made from scrap, usually junk pistons, lawnmowers and extrusion (screened porch rail).

The original project focused on a propane powered genset with heat recovery, however the price of propane nearly doubled since the original project was built, leading me to investigate the possibility of using waste oil products in a diesel engine. Propane is currently selling for approximately $3.00 per gallon while burner fuel is approximately $0.60 per gallon. The higher efficiency of the diesel engine, the heat recovery and the price of fuel may make a viable project for those in remote locations.

Waste oil products are often sold as burner or boiler fuel and may include vegetable oils, off spec diesel, jet fuel and used motor oil. Such oils must be filtered and may require heating before use, however sufficient information is included regarding the properties of oils for successful operation in such situations.

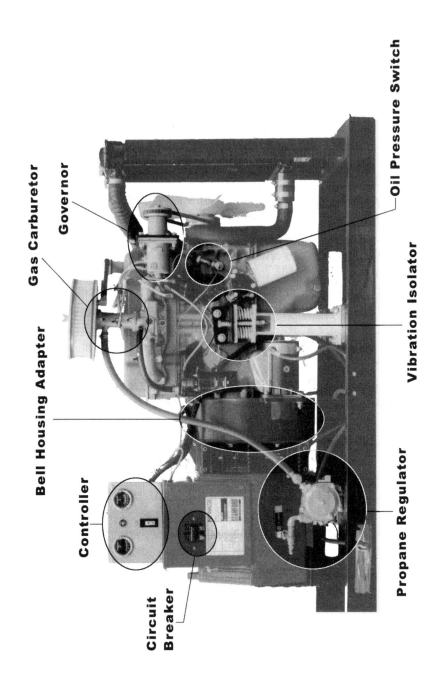

Oil Pressure Switch

Gas Carburetor

Governor

Vibration Isolator

Bell Housing Adapter

Controller

Propane Regulator

Circuit
Breaker

60

The generator heads used on these projects were both Western and Asian types. The Western types were of a higher quality and provide excellent power, however the Asian generators worked satisfactorily and were much cheaper. Details are found in the AC generator section.

PARTS OF A GENERATOR SET:

GOVERNOR: Maintains constant engine speed under various loads. It is essential for voltage and frequency regulation.

GAS CARBURETOR: Replaces the gasoline carburetor and allows operation on propane or natural gas.

BELL-HOUSING ADAPTER: Mates the generator frame to the engine.

CONTROLLER: Contains the start/stop switch, gages and engine protection device for low oil pressure and high temperature.

CIRCUIT BREAKER: Isolates the generator from the electrical buss. Protects the generator from overload.

PROPANE REGULATOR: Reduces tank pressure from 150+ psi to a few ounces.

VIBRATION ISOLATOR: Prevents transmission of engine vibration to the frame and foundation.

LAYOUT OF THE HEAT RECOVERY SYSTEM: The heat recovery system is based on two shell and tube heat exchangers. One exchanger recovers the exhaust heat. The other transfers the engine and exhaust heat to a heat sink or hot well. The exhaust heat exchanger is designed to accept engine jacket water so that the temperature of the exhaust gas will not drop below the dew point causing condensation and wet exhaust. However, because the exchanger is made from stainless steel, it does not matter if there is condensation in the heat exchanger. The cooling water flow may enter the exhaust heat exchanger and then flow to the engine. This may be a better flow path because the water entering the engine will be warmer.

The heat exchangers should be installed horizontally with the water connections on top so that they are always full of water. It is important that the thermostats are arranged so that there is always a flow of water through the heat exchanger, even upon starting when the engine is cold. The heat exchangers should be located below the level of the cylinder head so in the event of a leak, water will not drain back into the engine.

The original system design incorporated a header tank in a water-cooled exhaust manifold. The casting procedure is described in the casting section. The diesel revision does not yet have a water cooled manifold.

Cooling water is stored in a salvaged 500-gallon propane tank. It takes a little less than 4 hours to heat the water to 200° F when the genset is fully loaded. The heat exchangers were designed to operate with 160° F input water at which point the genset should be turned off or an additional cooling system employed. If a radiator is used, it should be twice the capacity of the typical radiator because it is responsible for nearly twice the heat input. I am currently experimenting with a pool water heater as a secondary heat sink. Due to the chlorine in pool water, this heat exchanger must also be made from stainless steel. Because the operating temperature is very low, it is of all welded construction with no expansion joints.

Water heaters consume 20% or more of the household electricity, a water-heater loop is added to the 500-gallon tank to preheat the water as much heat as possible before it enters the water heater. An electric water-heater is not a suitable load for an inverter therefore it is best run when the genset is operating or the water heater should be fired by propane.

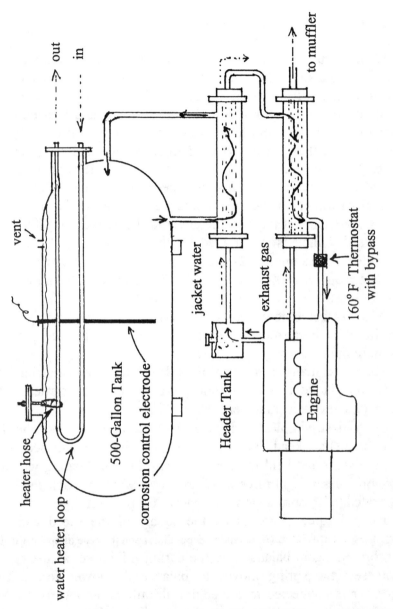

Drawing is schematic only. Actual location of parts will vary. The thermostat is located at the exhaust exchanger water outlet and has a small by pass for water flow even when starting cold. Additional pump will be required for the 500-gallon tank flow.

15. ENGINE GOVERNORS:

Both output frequency and voltage of an alternator vary with its rotating speed or RPM; therefore, engine generator sets require precise control of the speed of rotation. Such control requires an engine governor that provides both high accuracy and considerable force to operate the throttle. While there are many different types of governors, they can be divided into variable and fixed speed types. The cruise control on a car is an example of a variable governor. While it maintains the speed of an automobile at a particular setting, it is not permanently set to one single speed. An engine generator set requires the governor be permanently fixed to one speed, which is determined by the number of poles in the alternator. For example a 2-pole alternator requires 3600 rpm, a 4-pole set requires 1800 rpm, a 6 pole set requires 1200 rpm and so on.

It is important to note that the engine flywheel plays an important role in maintaining proper engine speed. See the flywheel section for more information.

There are many variations of both electrical and mechanical governors, however for this book, we are only concerned with droop type mechanical governors.

The oldest, simplest and yet one of the best governors is the centrifugal ball head. Two moveable weights are attached to a rotating shaft and held in place by a spring. The rotating weights develop a centrifugal force equal to mv^2/r, which is balanced by a calibrated stiff spring. As the speed of rotation increases, the centrifugal force increases by the square of the speed and the spring is compressed (or extended depending upon the type of spring) until the force is again balanced by the spring. Likewise, as the speed decreases, the spring moves to balance the lower force. The governor is connected to the engine throttle so as the flyweights move out, the throttle closes. As the flyweights move in, the throttle opens.

Antifriction bearings are used in governors to reduce the dead band or engine speed change without correction, however some friction is required in the system to prevent hunting. This friction is called damping and prevents the governor from perpetually surging and slowing around the target speed. In smaller governors, the viscosity of the lubricating oil provides sufficient damping without creating a deadband.

The force available to move the throttle is described by the ballhead scale. If a ball head has two flyweights and a .010 inch movement in one flyweight toe produces a force change of .2 pound, then the flyweight scale is 20 lb/in. Because there are two flyweights on the ball head, the ballhead scale would be 40 lb/in which would be typical for a good commercial governor.

The action of the ballhead is balanced by the speeder spring. By properly selecting the ratio of speeder spring to ballhead scale, a wide range of governor sensitivities can be had. If the scale of the speeder spring is *less* than that of the ballhead scale, then the governor will snap from one extreme position to the other being useless as a speed controller. However if the scale of the spring is appreciably greater than the ball scale, a stable system results.

Estimation of the required minimum spring scale is fairly simple. First, estimate the radii of the inner and outer weight orbits at the full

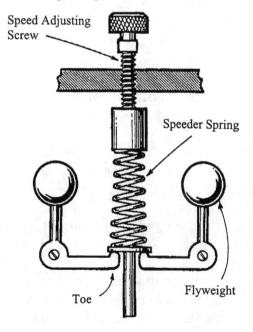

Speed Adjusting Screw

Speeder Spring

Toe

Flyweight

load and no load conditions. One weight is considered as the other one furnishes the reaction.

Calculation of Spring Scale:

r_1 =   radius of inner orbit
r_o =   radius of outer orbit
F_{ci} =   Force lb. at inner position
F_ω =   Force lb. at outer position
W =   Weight lb.
N_i =   Speed of governor at inner radius
N_o =   Speed of governor at outer radius

$$F_{ci} = 0.00034 \ Wr_iN_i^2$$

$$F_\omega = 0.00034 \ Wr_oN_o^2$$

The required sping scale equals:

$$(F_\omega - F_{ci}) / 12(r_o - r_i)$$

Which is the difference between the centrifugal force at the two positions divided by the difference in radius (spring compression) in inches.

Because spring scales are linear and the force of the ballhead varies with the square of the speed, governors for a wide range of operating speeds use conical springs. Conical springs have a variable spring scale. At low speeds all of the windings in the spring are active. At high speeds, the lighter turns close, leaving the smaller stiffer coils active. This type of spring is not required for our fixed speed governor.

Governors are also described as being drooping or isochronous. The speed decreases slightly as the load is applied with a droop type governor. The speed remains the same regardless of load with an isochronous governor. Droop type governors are used on small gensets and when paralleling multiple sets together.

Electric Governors:

Electric Governors have three parts: a speed sensing device, a device that compares the speed with a reference speed (voltage) and then outputs a signal and a device that actuates the throttle or fuel metering device relative to the input signal. The speed sensing device is usually a magnetic pickup that is mounted next to the gear teeth of the flywheel. The number of teeth on the gear determines the number of pulses per revolution of the flywheel. A 100 tooth gear turning at 1800 RPM would produce a frequency of 3000 Hz.

Magnetic Pickup Frequency = (number of teeth x RPM) / 60

The speed sensor coverts the frequency to a DC voltage relative to the speed of the engine. The voltage is compared to a reference voltage set by a potentiometer.

Governor Regulation:

Speed regulation in percent is calculated by:

100 x (no load speed – full load speed) / rated full load speed

Governor Droop in percent is calculated by:

100 x (no load speed – full load speed) / no load speed

16. MAKING A GOVERNOR:

A good commercial governor costs about $400.00, which is more than the rebuilt 4-cylinder engine we are using for this project. Luckily, a very good, high quality mechanical governor can be built mostly from scrap and costs about $15 to $20. The bearings and seals are purchased. Everything else is made in the home shop. The governor body is sand cast from scrap aluminum. The flyweights are machined from a scrap section of 1 ½-inch diameter steel rod. Most of the remaining parts are made from 1/8-inch sheet steel plate or small sections of steel rod.

Note that this is book #6 in the small foundry series, all of the sand casting techniques used here have been discussed in "Metal Casting volumes 1 & 2." This is not a particularly difficult casting project. Only topics specific to this casting project are discussed here. The "5 gallon bucket furnace" is used to make both the governor and propane carburetor.

Turning the Governor Pattern

After casting and machining a few governors, some minor changes in the patterns are noted below. These changes simplify both chucking of the castings for machining and accurate setting of the cores.

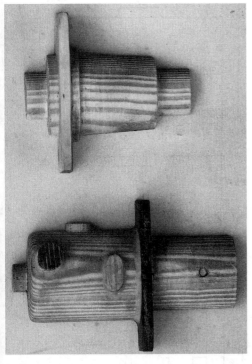

The two split patterns (left) for the governor body are made from yellow pine and turned in the lathe. There are large core prints on both front and rear patterns. Note that the *core* for the rear casting has 2 bearing bosses, one on each side. Although not seen in the photos, a flat should be added to both the core and the core print on the rear casting. This flat will prevent the core from shifting and keeps the internal and external bearing bosses aligned.

Seen in the drawings but not seen in the photos, the control arm bearing bosses should be extended down the sides of the rear casting. This change considerably simplifies chucking the casting in the lathe. The boss located on the top of the casting is the oil fill hole. It is drilled and tapped for a 1/8-inch pipe plug. Adding a boss and pipe plug to the bottom of the casting would simplify changing the oil in the finished governor.

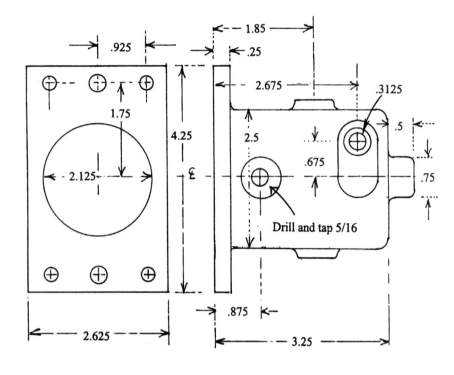

Drill and tap 5/16

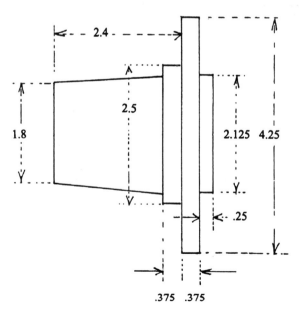

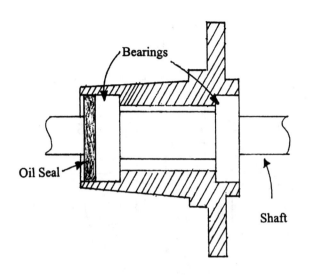

Front Cover - Cutaway

Bearing #	Seal #	ID -inch	OD -inch	Thick inch
R - 10	6370	0.625	1.375	0.2812
6003	6641	0.669	1.378	0.394
Seal #	6370	0.625		0.25
Seal #	6641	0.669		0.276

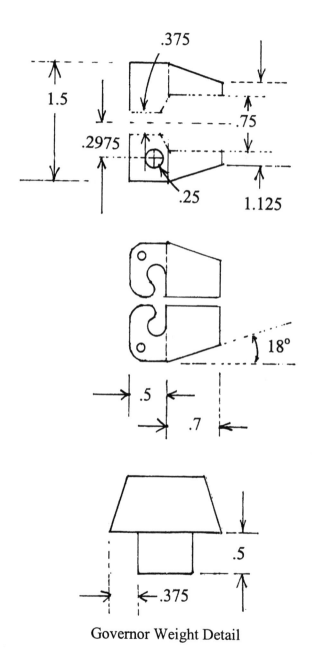

.375

1.5

.2975

.75

.25

1.125

18°

.5

.7

.5

.375

Governor Weight Detail

72

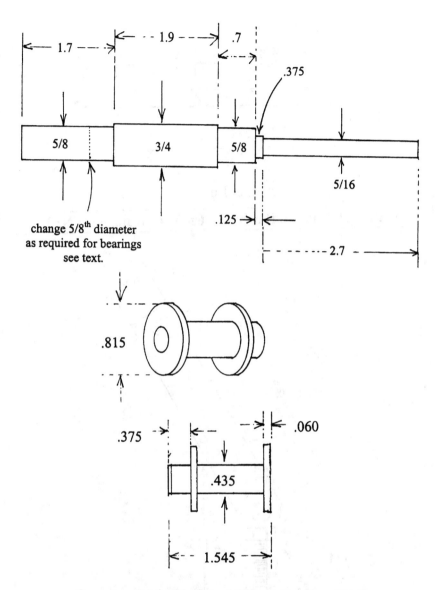

Governor Main Shaft and Thrust Bearing Shaft.

INA needle cage bearing number TC 815
INA Thrust Washer number TWA 815 2 pcs.
Available from MSC Supply (800) 645-7270

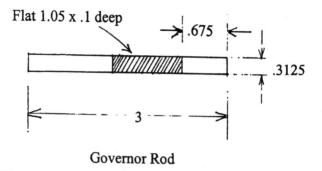

Flat 1.05 x .1 deep

.675

.3125

3

Governor Rod

Shaft Seal: Chicago Rawhide **CR# 3044** available at NAPA

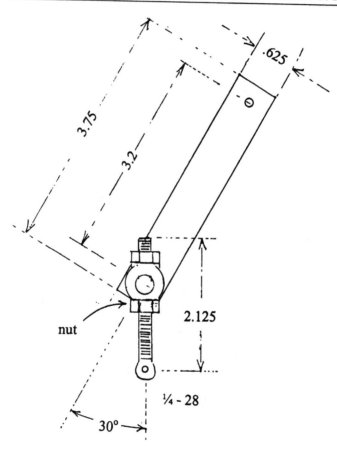

.625

3.75

3.2

2.125

nut

¼ - 28

30°

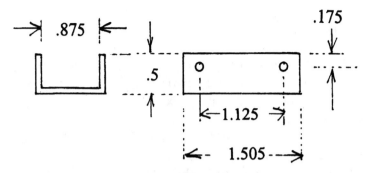

Flyweight Yoke

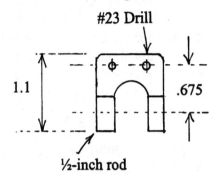

Speed Adjustment Arm

Making the Cores:

The cores are made from the typical home foundry mix of sand, wheat paste and molasses. One part molasses is added to eight or ten parts hot water. The water mixture is stirred until all of the molasses is dissolved.

In order to keep the dust down, splash a little molasses water in 100 to 120 mesh sand and mix it until the water is well distributed. Using a smearing action, stir in about 5% by volume dry wallpaper paste. Add a little molasses water and dry paste until a bond begins to develop. Continue to mix the core sand with a smearing motion until the mix is stiff enough for the cores hold together out of the core box. Mixing time is usually about 10-12 minutes. Insert a few sections of rebar tie wire to reinforce the cores as they are being rammed up in the core box. Bake the cores at 325° F. Using a file or piece of stiff wire, cut one or two vents opening at the end of the core print.

Mix a batch of core paste by adding flour to water and stirring until creamy. Glue the hot cores together and run a section of wire through the vents to be sure that they are not clogged with paste. Return the cores to the oven for a few minutes to dry. When cool, clean them up with a file or sandpaper.

Pouring the castings:

Ram the mold and pay attention to proper setting of the cores. The rear casting has a hanging core. If it shifts in the mold, it may be secured by pressing a few nails around the core print.

The castings are poured at approximately 1325° F and only a small riser is used to reduce shrinkage at the gate. This is a simple casting job and you should have no trouble with it.

Rough Governor Castings

Machining the Rear Casting:

Center the rear casting in the chuck using a surface gage. Face the front flange and bore it 2.125-inches diameter to accept the front housing. While in the lathe, chuck a ½-inch diameter end mill in the tailstock and cut a hole .550-inch deep in the rear bearing boss of the governor body. To assure proper alignment with the front surfaces of the casting, this operation must be performed while the casting is chucked for the surfacing and boring operations.

Locate the .3125-inch hole in the center of the side bearing boss, approximately 2.75-inches back from the front and exactly .675 up from the centerline of the governor. Drill 19/64-inch and continue through to the opposite boss but do not drill completely through the second boss. Stop approximately 1/8-inch before the outside surface. For a smooth bearing surface, finish the holes with a .3125-inch reamer. I may have used a .3135 reamer (one thousandth over). The material is soft and should work either way.

Centering the Casting

> Note that the governor could be made universal if the control arm shaft went through both sides and two seals were used. This would allow the arm to be moved from one side of the governor to the other.

Without moving the casting after drilling and reaming, use a ½-inch end mill to cut a relief for the shaft oil seal. The cut is .135-inch deep.

Locate the hole in the spring boss on the centerline of the casting and in the center of the boss. Drill and tap 5/16-24. While it is best to drill and tap a blind hole. I drilled through filled the hole with 5-minute epoxy when I inserted the screw. The epoxy both secures the screw and prevents oil from leaking around the threads.

Locate, drill 21/64-inch and tap the 1/8-inch 27 NPT pipe plug holes. Finally locate and drill the holes in the front flange.

A bronze bushing is used in the rear speeder shaft bearing. The material is ½ inch diameter SAE 660 bearing bronze. The finished bearing is .503 to.505-inch diameter, .5-inch long and drilled and reamed to .3125-.3135-inch for the shaft. Chamfer the ends to make pressing the bushing into the governor body easier. Turn down the end of a steel rod to make a tool to press the bushing into place.

Machining the Front Casting:

Checking the Bearing Bore

I used open standard sized R10 bearings in my governor, however the use of slightly larger metric bearings allows you make the bearing bosses on the shaft a little larger than the 5/8-inch diameter pulley shaft. This reduces the distance the bearing must be pressed during assembly. The drawings are dimensioned for the metric bearings.

Center the front casting in the lathe and cut the mating flange flat, leaving enough material for the .25-inch deep boss, 2.125-inches outer diameter. Bore the inner diameter of the flange 1.373-inches, .5-inches deep. This leaves a stop for the rear bearing. Move the boring bar 1.9-inches deeper (towards the front), and bore the remainder of the casting 1.373 inches in diameter for the front bearing and 1.375 for the seal. Move the casting to the mill to drill and tap 4 holes ¼-20. These screws hold the governor halves together. The 2 larger holes, 13/32-inch diameter, are for mounting the governor to the engine.

Making the Internal Parts:

The flyweights are cut from a section of 1 ½-inch diameter steel rod. Chuck the steel rod in the lathe and make a light cut down the length to insure that the finished weights will be symmetrical. Face off the end and drill a 3/8-inch diameter hole approximately 1.75-inches deep. Follow this with a .75 inch drill, .75-inches deep. Set the tool post to approximately 18° and cut taper on the end until

you reach a diameter of 1.125-inches. The taper will extend back approximately .7-inches from the end. This dimension will vary slightly because the diameter of the rod will be slightly less than 1.5 inches after cleaning up.

Left: splitting the weight

Move the rod to the mill and mount

80

it in a dividing head or some type of indexing fixture. Using a ½-inch end mill, make a cut .375-inches deep at the large end of the taper. Using the indexing fixture, turn the rod over and make an identical cut on the opposite side.

Split the weigh using a 5/16-inch end mill. Cut to the edge of

the taper. Using a ¼-inch end mill, drill through the flat .2975 off center and mill a slot across the rod. At each end of the cut, mill down .062 to leave a small toe in the center of the piece.

Using a slitting saw, or carefully using a hack saw, split the rod as seen in the photo. The slitting saw requires a pass down each side. Drill and ream to 1/8-inch the pivot holes in each side of the flyweights.

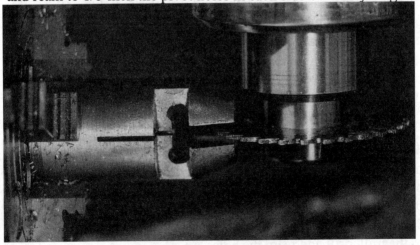

Sawing the Weights in Half

81

Cutting off the Weights in the Lathe

Return the part to the lathe and cut the weights off using a parting tool. Using a file, round the back edges of the weight and carefully round each toe. Weigh each weight to be sure that they are the same.

Make the flyweight yoke from a section of 1/8-inch plate. Give it a very light coat of paint on one side and scribe the layout lines using a caliper. Make the bends in the plate by clamping it in a vise and pounding it over with a hammer. Make the 90° bend as a series of small bends until it is flat against the vise jaw. Do not try to make the bend with own or two good whacks of a hammer. If plate will not fit in the vise jaws, cut a small hardwood or steel block that will fit inside the bend and hold it away from the jaws. The yoke is a precision assembly, work carefully. Move the yoke to the mill, drill and ream the holes.

The flyweights are held in place by 1/8-inch diameter water hardening drill rod. Cut the rods about .075 to .1 inch longer than required. The ends of the rod are expanded with a punch during assembly. Notice the prick punch holes in the photo. The flyweight yoke is welded or brazed to the finished speeder shaft. Be careful not to overheat the yoke or shaft, causing them to warp.

Flyweight Yoke Assembly

Turn the speeder shaft from water hardening drill rod. The fits are very important so accurate work here is essential. Polish the finished rod with #600 paper, being careful not to turn it undersized.

The thrust-bearing shaft fits over the speeder shaft and transmits the force from the flyweight toes to the control arm. Both the inner bore and the toe-face must be finished as smoothly as possible. The flyweight toes **must** move smoothly across the face and the whole assembly **must** slide smoothly on the shaft. In operation, the whole assembly spins with the speeder shaft. A needle thrust bearing rides between two ground thrust washers. This allows the spinning bearing shaft to transmit the horizontal movement to the governor control arm rod.

Thrust Bearing and Shaft

Note that the shaft has a groove for a wire retainer or snap ring. I did not use one in my governor, however it would surely prevent the bearing from ever slipping of the shaft.

The control arm yoke is made from a 1.1-inch length of 1-inch by 1/8-inch flat stock. You can also cut this from plate. The control arm yoke has section of ½-inch bar welded to it to provide a round bearing surface. A flat is cut on the rod so that it fits against the plate. Drill two ¼-inch holes in the plate behind the rod and weld the rod through the holes. Notch the assembly with a mill, however it may also be cut with a hand grinder. If the notch is not deep enough, this point may bind in the finished governor. A little trial and error fitting is required here.

Make the control arm rod from .3125-inch diameter water hardening drill rod. Cut a flat to accept the control arm yoke. Drill and tap #6-32. The control arm is assembled in the rear governor body. It fits over the speeder shaft and contacts the rear thrust washer.

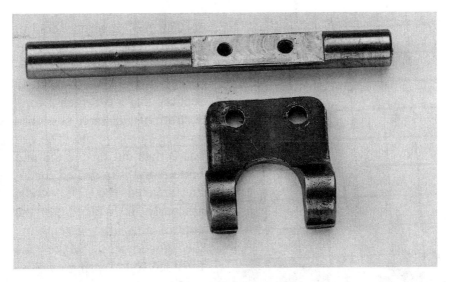

Control Arm Rod and Yoke

Rotating Flyweight Assembly

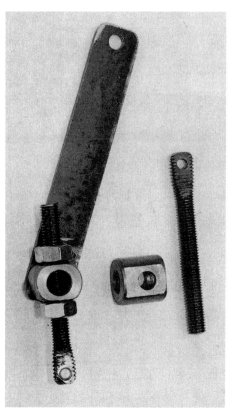

Cut the governor control arm from 1/8-inch steel. It is 3.75-inches long and .625-inches wide. Weld it to a section of 5/8-inch long section of 5/8-inch steel rod as seen in the photo. The rod has 2 flats cut on the top and bottom and it is drilled for ¼-28 threaded rod. The whole assembly is welded to the control arm after assembling the governor. Make the threaded rod "spring eye" by heating the end of the rod and pounding it flat with a hammer. Drill a 1/8-inch diameter hole for the end of the spring.

An adjustable lever attached to the side of the governor holds the opposite end of the spring.

Control arm

While the governor works very well with a small load on the generator, at very low loads the governor will hunt slightly. I have added a low speed stop, which is a 6/32 screw that strikes the control arm yoke. Tap the governor body for the screw but leave the last few threads imperfectly cut so that the screw slightly binds. I also cut a relief in the back of the governor for a nut to lock the stop in place. To adjust the stop, run the engine at no load and slowly advance the stop screw until the hunting just stops. Tighten the lock nut to prevent the setting from shifting. See the photo below.

Rear of Governor. Note the socket head cap screw used as a no-load stop.

Cut a gasket to fit between the front and rear covers. I used a section of cardboard, but would recommend that the gasket be cut from some proper gasket material available from an auto parts store. Oil soaks through a cardboard gasket and drips out over time. Fill the governor about 1/3 full of SAE 30 oil. In cold weather, thinner oil may be required.

Chuck the governor input shaft in the lathe and block up the governor body so that it does not spin. Run the governor for about a half an hour to break it in. Drain and replace the oil.

Governor Spring: See the chapter regarding winding a governor spring.

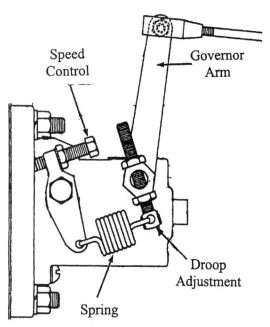

Speed Control

Governor Arm

Droop Adjustment

Spring

Make the governor to carburetor linkage from a section of threaded rod and two throttle ball joints available from auto parts stores such as NAPA, part number 2-685. Governor operation is affected by the lengths of the arm on the governor, the length of the arm on the carburetor and the length of the linkage between them. Pay close attention to the length of the carburetor arm.

GOVERNOR ADJUSTMENT: The speed is controlled by tightening or loosening the spring via the speed control bolt. The screw eye on the governor arm changes the sensitivity or droop of the governor with a longer screw being less sensitive. Too sensitive a setting causes the governor to hunt.

Begin adjustment with the hole in the droop adjustment screw being about 1-inch from the center of the control arm shaft. Using a multi meter with a frequency counter, start the genset and adjust the governor for 63 Hz. Load the genset to about half capacity and check the frequency. It should be around 61 Hz. Fully load the genset and check the frequency. It should be not be less than 58 to 59 Hz. For 5% regulation, there should not be more than 3 Hz difference between full load and no load. Changing one setting usually affects the others, making proper adjustment a trial and error affair. Continue adjusting the load, sensitivity, and speed until the sensitivity is close enough to provide good regulation around 60 Hz.

Governor RPM: The input shaft speed range of the governor affects both the force available to operate the throttle and sensitivity of the governor. Higher speed increases both providing tighter regulation. The speed range of the governor input shaft is controlled by the pulley size with smaller pulleys increasing the governor speed. My governor runs at 2850 RPM, which is 1.58 times the crankshaft speed. The crank pulley is 4.75-inches in diameter and the governor pulley is 3-inches in diameter. Governor speeds between 2850 and 3250 rpm should provide you with excellent results. For more information regarding pulley sizing, see the centrifuge section.

Pulley Drives: Finding an open pulley on the crankshaft may be a problem on some engines with few accessories. This is addressed in the chapter "Using Castings to Solve Problems."

GOVERNOR MOUNTING BRACKETS:

Most engines have accessory mounting holes on either the crankcase or cylinder head. The Ford engine has several flats with taped holes on the head. A mounting bracket made from ¼-inch steel plate holds the governor. A stiffening bar is made from ½-inch steel conduit. The ends are crushed flat in a vise and drilled for bolts. The bar fits between the top governor mounting bolt and the carburetor bolt on the manifold making a very rigid assembly.

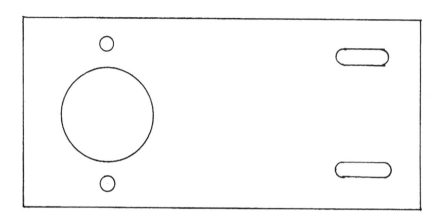

Bend ends at 90° as required.

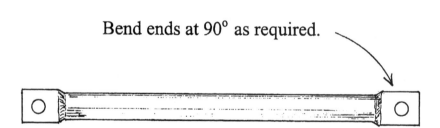

CARBURETOR LINKAGE: Threaded rod and ball joints are used between the governor and the carburetor. The NAPA number for the throttle ball joints is: # 2 - 685

17. SPRING WINDING:

Winding springs is not difficult and a useful skill for anyone working with machinery. Minimal tooling is required and it may be fabricated in a half an hour or less.

Springs are wound a mandrel held in the lathe. Spring wire is drawn through a tensioning fixture that is moved along the arbor as the spring is wound. For closed coils, the fixture may be fed by hand. If open coils are required, the fixture is mounted in the tool post and proper gearing is selected. My fixture is made from scrap oak wood and a half-inch bolt.

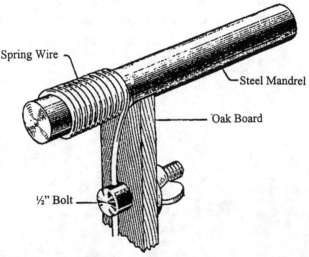

The tension of the feed and the closeness of the coils affect the final tension of the spring. Holding the fixture with a slightly backward angle seems to add a little pre-load to the spring making it a little stiffer. Spring winding is a trial and error process. It is best to wind a few practice springs with some soft wire such as rebar tie wire or copper. Be sure that you have enough free wire to complete the spring because if it comes out of the holder, it will spin around and make a painful and messy cut across your wrist.

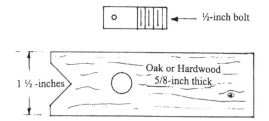

½-inch bolt

Oak or Hardwood
5/8-inch thick

1 ½ -inches

The mandrel is made from a ½-inch diameter steel rod and held the lathe chuck that is turned by hand while winding the spring. Upon releasing the tension after winding, the spring will expand off the mandrel. Mandrels have to be fitted and cut for springs that require an exact diameter. In order to have a clean bend for forming the end hooks of the spring, the mandrel has a .175-inch diameter wire through it

to form a smooth radius and hold the spring in place. The diameter of this wire is not critical as my wire is salvaged from a paper election sign. A hose clamp prevents the wire from slipping off of the mandrel. The most difficult part of spring winding is forming the end hooks. I have been more successful forming them over a small rod that is clamped in a vise. Cutting the wire with strong cutters is a job. A cut-off wheel may be a better solution.

The governor spring is wound from 15-gage carbon steel wire. Such wire is very inexpensive and available from tool suppliers such as MSC, among others. As stated, spring winding is a trial and error process. Usually three or four practice runs are made before the process is worked out and good springs result.

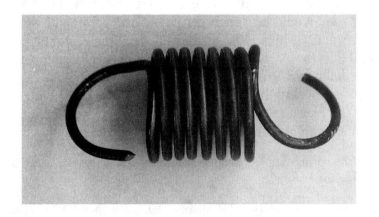

Completed Governor Spring

18. SPRING DESIGN:

Springs are used in the governor project and for vibration isolation, therefore a very brief description of spring principles is included here. More detailed information is available in books regarding machine or spring design and "Machinery's Handbook."

Springs with equal coil diameters (nonconical) are linear devices. This means that they deflect at a constant rate under a load, for example: if a 1 pound weight causes a deflection of 1 inch the *Spring Rate* would be 1 inch per pound and a 2 pound weight would case a deflection of 2 inches. A fish scale would be a common device based on this principle. To determine the spring rate of an unknown spring, measure the unloaded height of the spring with calipers to the nearest .001 inch. Add a known amount of weight and measure the loaded spring. Divide the deflection by the amount of weight to determine the deflection per pound.

Determine the spring rate for a valve spring so that it may be used as a vibration isolator:

The spring is approximately 1.235 inches OD, .8 inches ID, .177 inch wire diameter. The unloaded height is 1.98 inches.

A 25-pound weight is added to the top of the spring and the deflected height is measured to be 1.88 inches.

(1.980 −1.865) inches / 25 pounds = .0046 inch / pound

1 / (.0046 inch / pound) = 217 pounds per inch

The spring rate is 217 pounds per inch.

Spring behavior is governed by the spring material, the diameter of the spring wire, and both the diameter and number of coils. The ratio of spring diameter to wire diameter D/d also plays a role the spring stiffness. For general industrial springs, the D/d ratio is 8 to 10. For valve and clutch springs, 5 is common. Workable springs my have ratios between 5 and 14. Smaller spring coils make stiffer springs, likewise larger coils are softer. Increasing the number of coils lowers the spring rate. Springs are designed below the elastic limit of the material, this means that the usual working deflection will not cause a permanent deformation of the spring.

Springs are designed by assuming a mean diameter and safe working stress, after which the wire diameter is found by substitution in the stress equation. Several trials are usually required.

Assuming only torsion and round wire. The **spring stress** is:

$S_s = 8KFD_m / \pi d^3$

$F = \text{load} \quad D_m = \text{mean coil diameter} \quad d = \text{wire diameter}$

$K = (4C-1)/(4C-4) + 0.615/C, \quad \text{where } C = D/d$

The value of K may be take from the graph below:

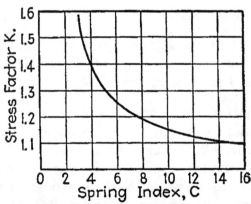

Deflection (*y*) of a coil spring made of round wire is given by:

$y = (n\pi D_m{}^2 S_s)/(KdG)$

n = number of effective coils, D_m = mean coil diameter, d = wire diameter, S_s = stress (see above), G= modulus of rigidity

G psi for various materials:

Carbon steel	11,400,000	Stainless Steel	10,000,000
Piano Wire	12,000,000	Phosphor Bronze	6,000,000

Designing Springs from Nomograms:

A quick and easy way to get a workable spring is to use the nomograms below. Note that they are based on the formulas above using carbon steel wire with a modulus of rigidity "G" of 12,000,000 psi. and 70% maximum load.

Design a spring using the nomograms: The governor project requires a spring rate of 18 to 20 pounds per inch. The spring should be between .625 to .75 in diameter.

1. Find the wire size: Go to nomogram #1 and select a coil diameter of .7-inch. Draw a horizontal line at this point. Select a load size of 19 pounds and draw a vertical line from that point. The two lines intersect between #16 and #14 gage wire size. Select #15 gage wire.

2. Determine the D/d ratio: Find #15 wire gage in the wire table and locate a diameter of .072. Divide the coil diameter by the wire diameter. The result is D/d = 9.7.

3. Find the number of coils: Go to nomogram #2 and locate a D/d ratio of 9.7. Set a ruler down at this point and locate the #15 wire gage on the S.W.G. scale. Draw a line across the reference line between these two values. Go to the spring rate scale and find 19 pounds. Set a ruler down at this point and cross the point marked on the reference line. Find that the number of active coils is approximately 7.

The governor spring will be made of 15 gage wire, have 7 coils and an outside diameter of approximately .7 inch. *

*Note that different combinations of wire and coil sizes may be used to make a spring of similar rate. Actual spring rate will also be affected by variables in the hand-winding process.

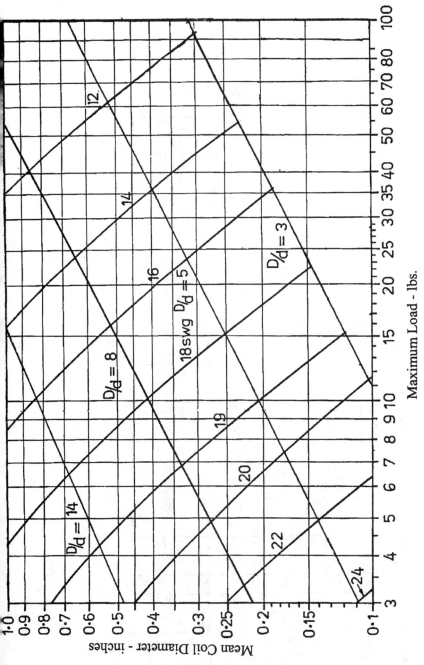

Spring Design Nomogram #1

97

Steel Wire Gages
Inches

Gage Number	US Steel Wire Gage	Standard Wire Gage	Music Wire
6	0.1920	0.1920	0.016
7	0.1770	0.1760	0.018
8	0.1620	0.1600	0.020
9	0.1483	0.1440	0.022
10	0.1350	0.1280	0.024
11	0.1205	0.1160	0.026
12	0.1055	0.0140	0.029
13	0.0915	0.0920	0.031
14	0.0800	0.0800	0.033
15	0.0720	0.0720	0.035
16	0.0625	0.0640	0.037
17	0.0540	0.0560	0.390
18	0.0475	0.0480	0.041
19	0.0410	0.0400	0.043
20	0.0348	0.0360	0.045
21	0.0317	0.0320	0.047
22	0.0286	0.0280	0.049
23	0.0258	0.0240	0.051
24	0.0230	0.0220	0.055
25	0.0204	0.0200	0.059
26	0.0181	0.0180	0.063
27	0.0173	0.0164	0.067
28	0.0162	0.0148	0.071
29	0.0150	0.0136	0.075
30	0.0140	0.0124	0.080
31	0.0132	0.0116	0.085
32	0.0128	0.0108	0.090

*Spring Nomograms adapted from Tubal Cain.

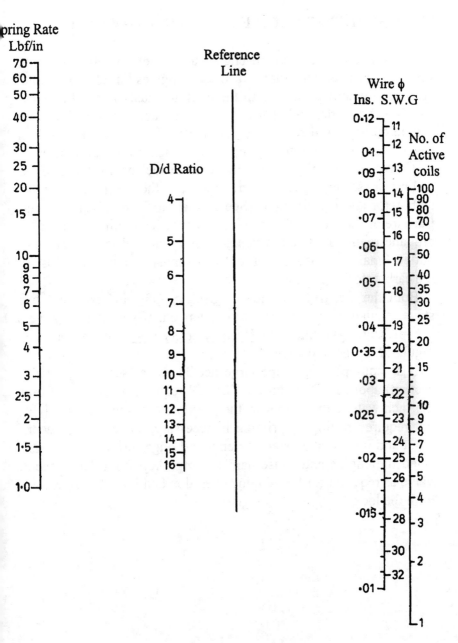

Spring Design Nomogram #2

19. CONVERTING ENGINES TO BIFUEL OPERATION:

Conversion of a gasoline engine to bi-fuel operation is quite simple. Gaseous operation of diesel engines is also possible, however the amount of gas in the inlet air must always be below the "limits of inflammability of the fuel gas used" or knocking and engine damage will result. Typically, diesel engines ingest a lean gas mixture that is ignited by injecting a small pilot charge of oil.

Conversion of a gasoline engine to bi-fuel operation involves adding a second gas venturi between the original gasoline carburetor and the air filter. (See the venturi in the "building Gas Carburetor Section). The venturi is attached with a hose to a propane regulator. A large needle valve is added to the regulator outlet. Finally, a gasoline cut off valve must be added as close to the carburetor as possible.

Both the Beam T-60 and the Century G85 are excellent two stage regulators for this type of conversion. If you already have a low-pressure gas line, the Beam Garretson - KN low-pressure regulator is a good choice.

If so equipped, open the large needle valve (see the gas valve section) about 2 - 2 ½ turns. The KN requires a vacuum of only .25 inch water column to start the gas flow and engines usually do not require priming. If priming is necessary, use a 1 or 2 second burst and immediately start the engine. Longer priming will likely flood the engine and cause hard starting. Adjust the large needle valve as required for best engine operation and lock the adjustment with the locknut.

20. COMMERCIAL GAS CARBURETORS:

Gaseous fuel systems are two part devices containing both a carburetor and regulator. Gas carburetors require a device to reduce and control the gas pressure from the LP tank to the pressure required by the engine. The carburetor functions as both a mixing device and to control the amount of air entering the engine. The vacuum generated in the carburetor venturi activates the regulator to increase or decrease the gas flow as dictated by the load on the engine.

There are two types of regulators, single stage or low-pressure types and two stage units containing both a high and low-pressure regulator. The first stage or high-pressure regulator reduces the gas pressure from tank pressure or about 250 psi to about 4 psi from which the secondary or low-pressure regulator controls the actual flow to the engine. High-pressure regulators usually have some means of heating, either by air or hot water, to vaporize liquid

propane. The Beam T-60 has a port to circulate the engine's jacket water or exhaust to provide heating while the Century G-85 depends upon fins as seen in the photo. (**Left:** Beam model T- 60 two stage regulator.)

Air heated units should be placed in the path of the radiator air discharge if practical.

101

Regulators also have a manual or electric primer that allows you inject a small burst of fuel gas into the carburetor, reducing the cranking time. Electric primers are typically used for remote start systems.

All regulators used on engines located inside a building or garage must have a positive shut off valve to assure that the flow of fuel will be cut off should the engine fail while unattended.

The positive shut off valve is an inexpensive option available on the regulators. Usually, the valve senses the manifold vacuum of a running engine.

(**Left:** The century G85 regulator front and back view. This regulator has an electric primer.)

Gas carburetors and regulators are available from several sources such as Beam, Century and IMPCO. Beam manufactures both the "KN" low-pressure regulator and the "T-60" which is a two-stage unit. The IMPCO AHR50D regulator is a direct replacement for the Century G85.

 IMPCO series #55 carburetors **(left)** may be mounted in the updraft, sidedraft or downdraft positions on engines from 1.5 to 70 hp. The maximum flow rate is 115 cfm. They are available in SAE flange sizes ¾ -inch, 1 inch and 1 ¼- inch. The 55 series gas carburetors have a self-contained secondary regulator and work well when paired with the Century G85 regulator.

The gas regulator is connected to the carburetor with a suitable rubber hose. The hose must be stiff enough not to collapse under vacuum and be propane resistant. Good quality automotive heater hose or NAPA multi-hose works well for this application.

Capacities of Small LP Gas Regulators:

Beam T50 unheated – gas withdrawal only   40 hp
      T60 Liquid Withdrawal   60hp, suffix A=primer, H= exhaust heat

Century G 85 Air heated, 50 hp

IMPCO Carburetor Part Numbers:

¾ - inch SAE Flange, 1 ½ inch air cleaner inlet, = CA55-12
1 - inch SAE Flange, 1 ½ inch air cleaner inlet, = CA55-14
1 ¼-inch SAE Flange, 2 1/16 inch air cleaner inlet = CA55-508

Beam Regulator Approximate Cost, May 2006: $65 - $85
Century Regulator Approximate Cost, May 2006: $95

IMPCO Carburetor, approximate cost May 2006: $125

21. DESIGNING GASEOUS FUEL CARBURETORS:

Making workable gas carburetors for generator systems is a surprisingly simple affair. The most simple gas carburetor consists of a throttle (butterfly), venturi and a low-pressure regulator. The main fuel-metering device is the low-pressure or secondary gas regulator. This regulator is activated by vacuum. As the vacuum increases in the venturi, the regulator diaphragm opens a needle valve, which allows more gas to flow. Because generators are run at constant speed, they do not require an idle system; however variable speed gas carburetors have a preset idle flow that is attached to a second port located after the carburetor butterfly valve.

The secondary regulator may be a stand-alone type such as the Beam "KN" model regulator or it may be contained in a dual regulator. Such regulators have both the high-pressure regulator and the low-pressure regulator contained in one device. The Beam model 60 is a dual regulator designed for LP conversion of engines 60 horsepower or less.

LP gas and natural gas systems are similar, the difference being natural gas systems require a higher volume of gas flow and the gas is usually supplied from a low pressure pipeline. LP gas is usually supplied from a tank from which the pressure varies with the temperature. The high-pressure regulator feeds the secondary regulator at a constant pressure regardless of tank pressure or flow.

DETERMINING VENTURI SIZE:

If the engine already has a gasoline carburetor, make the venturi the same size as the one in the existing carburetor. If the engine does not have a carburetor or has a fuel injection system, then size the venturi according to the airflow into the engine.

By making a few simplifying assumptions and not going into the fluid dynamics of carburetor flow, a workable system may be designed as described below.

1. *Assume steady flow through the venturi*. When 4 cylinders are connected to a venturi, this is approximately the situation. However when less than four cylinders are used, the flow is intermittent or pulsating. Pulsation is maximum for a single cylinder engine because intake flow occurs only for one half revolution for every two revolutions of a crankshaft. Interruption of the flow through the carburetor of a single cylinder engine causes a blowback of fuel and air through the venturi.

2. *Assume the maximum flow **rate** as if four cylinders were used.* (Although this flow will only occur during one half revolution every two revolutions of the crankshaft in a single cylinder engine.) Therefore a single cylinder engine will have the same maximum flow **rate** as if 3 or more cylinders were used.

3. Assume a maximum flow rate of approximately 82 feet per second through the venturi. While this is not an absolute, it is an average value seen on several industrial engines.

Example:

Size a carburetor venturi for a four-cylinder 1600 cc engine running at 1800 RPM. The bore of the engine is 3.188 inches and the stroke is 3.056 inches.

> There are several ways to determine the single cylinder displacement. I am showing some unit conversion here. Work in whatever system you are comfortable with.

Method 1: Divide the engine size by the number of cylinders.

1600 / 4 cylinders = 400 cc per cylinder.
Convert 400 cc to cubic inches:
$400cc / 2.54(cm/inch)^3 = 400cc / 16.39 = 24.41$ cubic inches

Method 2: displacement $= .7854$ (bore^2) x (stroke)

$.7854 (3.188^2)$ x $(3.056) = 24.39$ cubic inches.

Next, find the volume flow per second at 1800 rpm. Because the intake flow only occurs on every other revolution, divide the RPM by 2 for 900 intake strokes per minute. There are 60 seconds in a minute:

Flow $= (24.39$ inches^3 x 900 strokes per min.$) / 60$ sec. per min.

Flow $= 360$ cubic inches per second

Converting the maximum flow **rate** of 82 feet per sec. to inches:

82 feet per second x 12 inches per foot $= 984$ inches per second

Find the required venturi area for the maximum flow rate:

360 in^3 per second $/ 984$ in per second $= .366$ in^2

Find the diameter of the venturi:

Diameter $= \sqrt{(\text{area} / .7854)}$

Diameter of the venturi $= \sqrt{(.366 \text{ in}^2 / .7854)}$

Diameter of the venturi $= .683$ inches

An 11/16ths drill (.687) is pretty close to the required diameter and will be used for the venturi.

22. BUILDING A GAS CARBURETOR AND ADAPTORS:

Gas carburetors are easy to build, requiring only one casting, a butterfly valve and a venturi. Gas adapters fit in front of a standard gasoline carburetor and require only a venturi.

The gas adapter above consists of the venturi (top) and the sleeve (bottom). The needle valve (center row) attaches to the regulator outlet and is used to turn down the gas flow as required.

Gas adapters do not require castings and may be turned from round stock. The adapter mounts between the air filter and the gasoline carburetor. The vacuum in the venturi acts on a diaphragm in the gas regulator controlling the gas flow, making this an extremely simple device. Venturi construction is detailed in the following carburetor section.

The gas carburetor is made from a split pattern. A baked sand core1-1/8-inches in diameter by 6 inches long forms the inside of the casting. Standard SAE flange dimensions are used on the top and bottom flanges. These dimensions may be changed as required for your particular application. See the table on the next page.

Gas Carburetor Casting (center) and Patterns

Bosses are added for the gas hose-barb at the top and for the throttle bearings on the sides. A feature not included on this prototype casting is a throttle stop. While my carburetor functions well without it, in hindsight it would be a good thing to have. A small tab tapped for an adjustment screw would be fine.

Make the venturi from a section of steel rod, however brass or aluminum would also work well or better. Chuck the stock in the lathe, center drill it, then and step drill it to11/16ths-inch. Set the tool post to approximately 7.5 degrees and bore the taper. Cut the relief for the gas flow then saw the 1/16th-inch wide slots for the gas at the base of the inlet radius. Return the part to the lathe, and trim the part 1.125-inch long. Round the front of the venturi with a

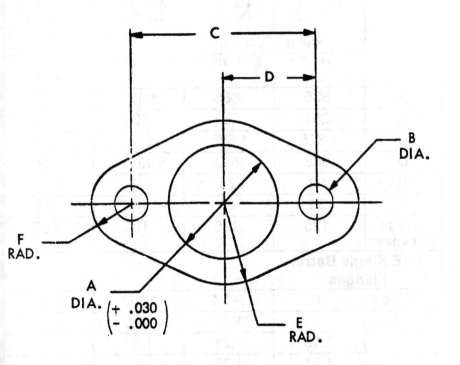

109

SAE Tapped and Flanged Connections			
Engine Horsepower Per Cylinder			
	1 to 4 Hp 75 to 3 kW	4 to 8 Hp 3 to 6 kW	8 to 16 Hp 6 to 12 kW	16 to 20 Hp 12 to 15 kW
Exhaust Port Tap			
	1/2-14 NPSF	3/4-14 NSPF	1 - 11-1/2 NSPF	1-1/4 - 11-1/2 NSPF
A	.635	.885	1.135	1.385
B	.281	.281	.303	.343
C	1.375	1.625	2.00	2.25
D	.688	.812	1.00	1.125
E	.62	.75	.88	1.00
F	.31	.31	.44	.44
Flange Thickness	.125	.125	.188	.188
SAE Single Barrel Flanges			
SAE Size	1	1-1/4	1-1/2	1-3/4
Bore Size	1	1.19	1.50	1.94
	1.19	1.21	1.51	
	1.21	1.25	1.56	
	1.25	1.44	1.69	
	1.31	1.50		
E Outer Radius	1.09	1.15	1.25	1.41
C Bolt Centers	2.38	2.69	2.94	3.31

file as seen in the flange drilling photo. Smooth the inside of the venturi with a piece of sandpaper. Later, the venturi is secured in the carburetor body with a set-screw.

The butterfly valve or throttle plate is cut from a section of 20-gage sheet steel. The throttle rod is cut from a section of ¼-inch diameter brass rod, which machines very easily. Drill rod would be the best; however drill rod is more difficult to machine. Locate and cut a flat on the rod using an end mill or, if you are working with brass, you may cut the flat with a file. Drill and tap the holes for the throttle plate #6-32.

Boring the Carburetor Body

The throttle plate or butterfly is NOT made to the diameter of the carburetor bore. It is cut .030 larger in diameter and the sides next to the rod bushings are filed or ground to .005 smaller in diameter than the carburetor bore as seen in the drawing. Using a #23 drill, locate and drill the holes in the throttle plate .050 off-center. This stops the throttle and prevents it from spinning around backwards in the carburetor bore.

The small arm for the throttle linkage may be cast, however I cut and welded a strip of 1/8-inch thick strap to a section of ½-inch

111

diameter rod. The assembly is drilled after welding and then a #6-32 set screw is added to secure it to the throttle shaft. The center of the throttle linkage is located 0.55 inches off center from the center of the throttle shaft.

Left: Rounding the front edge of the venturi by using a round file while turningthe venturi in the lathe. (Note that the photo is not to scale. This is a 2" diameter venturi.)

Left: Slitting the venturi by using a 1/16-inch slitting saw in the mill. (to scale)

Reaming the Throttle Bushings.
The carburetor casting is clamped to an angle plate for the drilling and reaming operations. I did not drill completely through the bottom bushing, making a blind hole.

Drilling the Carburetor Flange.

Note the position of the slot in the venturi and the set-screw that is holding the venturi in place.

Test fit the throttle shaft into the carburetor body and attach the throttle plate. When you are satisfied that the parts fit and work properly, coat the threads of the #6/32 screws with locktight or 5 minute epoxy and assemble the butterfly to the shaft. Oil the bearings and install the throttle plate. Drill, tap and install a hose bib at the gas inlet boss.

Tapping the Flat on the Throttle Shaft.

Propane hose: The connecting hose between the regulator and the carburetor must be heavy enough to resist collapsing under vacuum. NAPA sells a red colored, 250psi "Multi-Duty Hose," which is well suited for this application. Good quality automotive heater hose may also work.

Completed Gas Carburetor and Linkage
Note the Sand Cast Air Cleaner Plates.

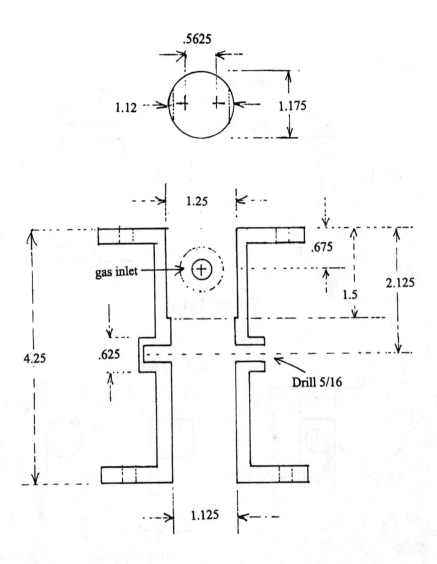

Carbruretor Body and Throttle Plate

Note: gas hose boss is ¾ -inch diameter. Make flanges per application.

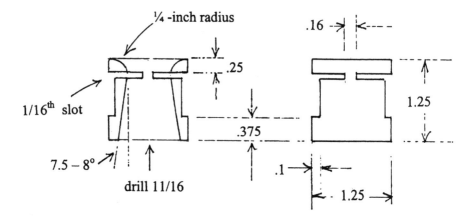

Venturi Detail

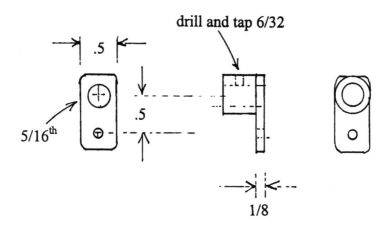

Carburetor Linkage Arm

118

AIR CLEANER PLATE:

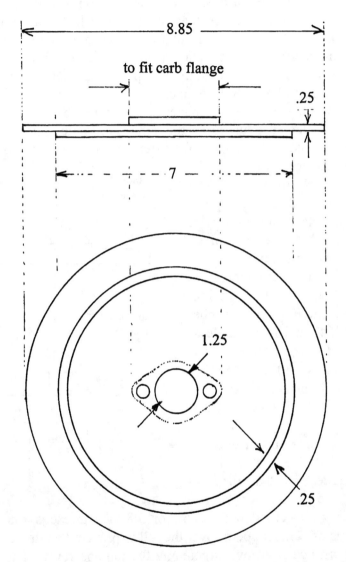

8.85

to fit carb flange

.25

7

1.25

.25

Plate Detail
Cast Aluminum

A fully covered air cleaner is probably better, especially in rainy weather but I wanted a quick and easy solution to my air cleaner problem. I cast two plates that require very little finishing and work quite well. Another rim or ring at the outer edge could be added and a strip of sheet metal attached for a more enclosed unit. The flange size is determined by your particular application. I used 1¼- inch single barrel SAE flange.

The air cleaner pattern is made from Masonite because it is both cheap and flat. It is turned on the lathe. The same pattern is used for the top and bottom. The only difference between the two plates is the center hole for the carburetor. I made the SAE flange as a detachable piece held in place by two wire brads so that I could ram up the mold on a flat pattern.

The air cleaner element fits an 86 Camero with a 2.5-liter engine. The Fram part number is CA3647 and is available as Walmart. The NAPA part number is 6036. The air cleaner dimensions are: H= 2.489   OD = 8.466   ID = 6.968

When the air cleaner is assembled, the plates should be marked so that they go back the same way. The assembly is secured with 5/16[ths] or 3/8-inch threaded rod. Short sections of ¼ inch pipe fit over the rod to prevent the element from being crushed from over tightening the thumbscrews. The inside diameter of ¼ inch pipe = .364-inch. You will have to run a 3/8-inch drill bit through it.

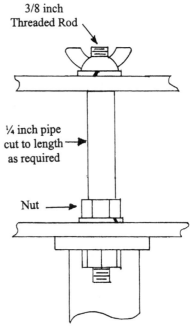

3/8 inch
Threaded Rod

¼ inch pipe
cut to length
as required

Nut

23. MAKING THE GAS VALVE:

The gas valve fits between the regulator and the carburetor allowing you to reduce the gas flow as required for best performance. It is a simple device consisting of a tapered screw that closes the opening to the hose barb. A locking nut secures the screw after proper adjustment. The valve described here works well for 12 kW and smaller generator sets. The valve may be scaled for larger sets. The only critical feature of the valve body is that it must be drilled deep enough for the adjustment screw to contact the inside surface of the hose barb or pipe nipple, depending upon orientation. Note that the adjustment screw may be positioned to close off the either the pipe nipple or the hose barb, whichever is most convenient for your situation.

Above: adjustable gas valve

I have made several versions of the valve, the drawings feature a valve made from 1-inch diameter brass rod, however steel or aluminum would also work. Brass is both easy to work and non-sparking. The screw is made from a section of brass rod with both the threads and the taper being cut in the lathe. Stock bolts or threaded rod may be substituted to make the job quicker and easier.

Bolts and threaded rods are easily clamped for machining by using a fixture made from a split nut. Cut through ONE side of a nut, then run a tap through it to clean any burrs off the threads. Run the nut down the threaded rod as needed to hold it in the lathe. As the lathe chuck tightens on the nut, it will clamp down firmly on the threaded shaft without damaging the threads. The split nut is also used to hold the screw to cut the slot at the top. While I used a slitting saw to cut the screw head slot, there is no reason that a hack saw would not work.

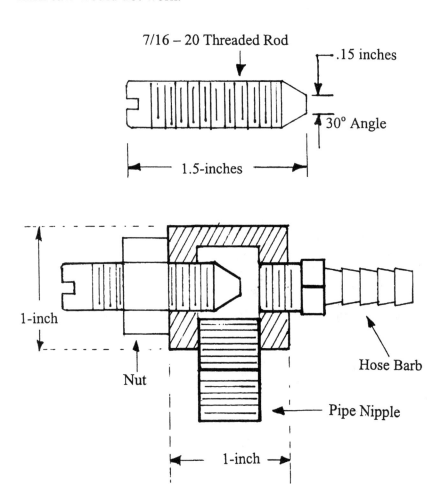

Gas Valve Cross Section

24. ENGINE FLYWHEELS:

Proper regulation of engine speed requires both a governor and an adequately sized flywheel. The power delivered by a piston engine fluctuates over a wide range during each revolution of the crankshaft. The flywheel acts as an "intra-revolution" governor by absorbing the energy delivered by the piston and releasing it during the remainder of the cycle. Therefore the function of the flywheel is to store up and restore the fluctuations in energy given to or taken from an engine, thus keeping the angular velocity of rotation approximately constant. Since the flywheel is absorbing and delivering energy, the speed varies during each revolution, displacing the flywheel ahead of and behind its mean position. The mean position being the position it would occupy if the angular speed remained constant.

Constant angular speed is important in engine generator sets to prevent lighting flicker, among other things. Using a 60-watt light bulb at 115 volts, flicker is just apparent with a voltage variation of 0.8% or about .92 volt. Proper voltage regulation requires that the maximum angular displacement be limited to 2-1/2 to 3 electrical degrees. Heavy flywheels also reduce the voltage drop during the application of large electrical loads such as motor starting.

Unfortunately for the reader, calculation of the required flywheel mass and dimensions to meet this standard involves tedious graphical or numerical integration of the engines time torque curve. Such calculations are described in machine design text and therefore not included here. Our discussion will involve a few general guidelines regarding flywheel size and safety.

Diesel engines, due to their much higher compression ratios, require much larger flywheels than to similar capacity gasoline engines. A greater number of engine cylinders reduces the required size of the flywheel, one and two cylinder engines having poor regulation. Four cylinder engines are good and inline six cylinder engines are very good. Clutch type automotive type flywheels are generally too light for good regulation and automatic transmission models are of no use what so ever. I have doubled the weight of the automotive flywheel that came on the Ford engine with good results. I would suspect that increasing the automotive flywheel weight by a factor of 2 or 3 should provide you with good to excellent results.

Stresses in Flywheels:

As a flywheel spins, centrifugal force attempts to pull it apart. The force is similar to that found in a thick walled cylinder under internal pressure. The stress is the greatest at the inner diameter (center hole) of the flywheel and it is at this point where cracks occur. Typically, when run too fast, flywheels explode with catastrophic results. Because the forces causing the stress are directly related to the rotational speed, there will always be a speed at which a flywheel explodes, regardless of the thickness of the material used. Therefore, a maximum safe speed should be calculated when designing a flywheel. Although there are more accurate methods, an easy way to estimate the maximum safe speed of a flywheel is taken from the speed of rupture.

For a given material, the speed of rupture is expressed by:

$$V = 96\sqrt{S/W}$$

V= rim velocity of the wheel in feet per minute,   S= tensile strength of material, W= weight in pounds of 1 cubic inch of the material

W=.26 for cast iron   W= .28 for steel

S cast iron = by class, such as class 20 = 20,000 psi

S steel = 60,000 psi

---

*Factor of safety = 10 Divide the tensile strength of the material by the factor of safety

---

Example: A 10.5 inch diameter class 20 cast iron flywheel is proposed for a generator set running at 1800 RPM. Determine if this is a safe speed for this diameter wheel.

1. Find the rim velocity of the wheel at 1800 rpm:

   D = 10.5-inches / 12-inches per foot = .875 foot

   Circumference of flywheel = πD π= 3.1416

   3.1416 x .875 = 2.75 feet

   Rim velocity V = RPM x Circumference

   V = 1800 x 2.75 feet = 4950 feet per minute

2. Estimate velocity of rupture:

   Class 20 cast iron has a tensile strength of 20,000 psi.

   Using a safety factor of 10: 20,000 psi / 10 = 2000 psi

$$V = 96\sqrt{2000 / .26}$$

$$V = 8420 \text{ feet per minute}$$

4950 feet per minute is less than the rupture velocity of 8420 feet per minute so the flywheel should be safe for this application.

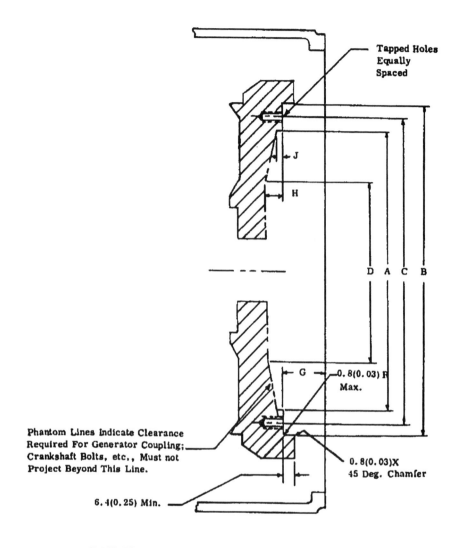

SAE Flywheel for Single Bearing Generators

SAE Flywheels for Single Bearing Engine Mounted Generators

Nominal Clutch Diameter	A	B	C	D	E	F	G	H	J	Number Tapped Holes	Size
6 1/2	7.25	8.500	7.875	5.00	2.81	2.50	1.19	0.50	0.38	6	5/16 18
7 1/2	8.12	9.500	8.750	NA	2.81	2.50	1.19	0.50	0.50	8	5/16 18
8	8.88	10.375	9.625	NA	3.94	3.00	2.44	0.50	0.50	6	3/8 16
10	10.88	12.375	11.625	7.75	3.94	3.00	2.12	0.62	0.50	8	3/8 16
11 1/2	12.38	13.875	13.125	8.00	3.94	NA	1.56	1.12	0.88	8	3/8 16
14	16.12	18.375	17.250	8.75	3.94	4.00	1.00	1.12	0.88	8	1/2 13
16	18.12	20.375	19.250	10.00	3.94	4.12	0.62	1.12	0.88	8	1/2 13
18	19.62	25.500	21.375	NA	3.94	4.12	0.62	1.25	1.25	6	5/8 11
21	23.00	26.500	25.250	NA	3.94	5.75	0	1.25	1.25	12	5/8 11
24	25.38	28.875	27.250	NA	3.94	5.75	0	1.25	1.25	12	3/4 10

24. MAKING FLYWHEELS AND COUPLINGS:

There are a number of ways to solve the flywheel and coupling problem. A complete new flywheel may be cast and machined, a doweled coupling disc may be cast or a doweled "doughnut" ring may be machined from steel plate to fit between the flywheel and generator coupling. I cast a disc to fit between the flywheel and the generator. The disc is located by dowel pins in the holes originally used to locate the clutch. The addition of the disk doubles the weight of the original flywheel. While there may be some concern regarding the additional loading on the rear main bearing, the clutch and pressure plate assembly are very nearly the weight of the additional disk.

Considering the time required to cast and machine the bolt on disk, making a completely new flywheel would not be any more difficult or time consuming. A split pattern may be used. One side being the crankshaft side and the other being the generator coupling side. The only other consideration would be the ring gear, which is shrunk onto the flywheel. Uniformly heating it with a torch expands the ring enough to easily remove it from the original flywheel. Carefully measure the diameter of the existing flywheel under the ring gear. You will have to cut the new flywheel with a ring of this diameter and location from the end of the crankshaft. After machining your new flywheel, heat the ring gear and slip it over the new flywheel. Upon cooling, it will grip the new flywheel securely.

My iron castings are made from cupola iron. (A cupola is a type of furnace.) Cupola melting increases the amount of carbon in the iron and may change its properties. The class of iron is directly related to the type of iron or steel charged into the furnace. Generally speaking, the higher the carbon content, the lower the class of iron. For higher tensile strength, I use more steel in the charge; usually 50% scrap iron and 50% steel adding a little silicon to the ladle for easier machining. Because the castings often have a hard "skin" of white iron (combined carbon) I will grind the entire surface that is to be machined to cut through this skin. A hand grinder works well for this job and the castings usually cut like butter if this skin is removed. Failure to remove this skin makes for difficult machining and very rapid tool wear.

The metallurgy of iron is described in *"Metal Casting Volume 2"* Cupola melting is described in *"Iron melting Cupola Furnaces for the Small Foundry."*

Making the Pattern: The best circular patterns are made from built up pie shaped sections, similar to those used in the flywheel-housing pattern. However, to simplify matters, I made my pattern from glued up discs of plywood and Masonite. The built up edge was made from six strips similar to the ones seen in the flywheel-housing chapter. The wooden patterns were glued together and turned on the lathe. A heavy layer of Bondo was then added to the inside surface. When dry, it was cut to form the taper between the

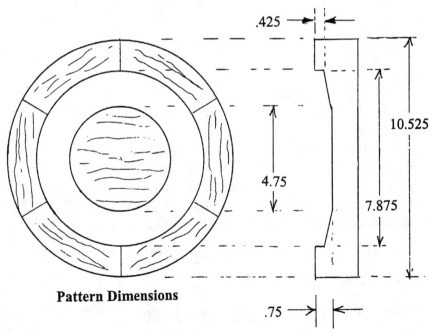

Pattern Dimensions

outer ring and inner surface of the flywheel. Both the outer diameter and inner diameter of the ring are tapered 5° to form the draft required to remove the pattern from the sand.

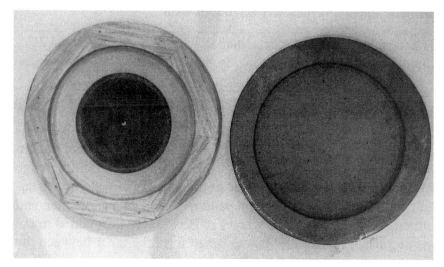

Wood Pattern and Flywheel Casting

These dimensions may or may not be suitable for your application. This flywheel coupling was machined for a SAE # 7 ½ coupling. The total thickness of this pattern is 1.375-inches. If you were making a completely new flywheel, I would add another inch to the thickness.

The weight of the flywheel coupling may be approximated by calculating the weight of the iron disc and subtracting the weight of the cut away portion. Assume iron weighs .26 pound per cubic inch. Volume of a cylinder = $.7854D^2L$, OD = 10.375, ID = 7.875 Thickness of wheel = 1.375, Thickness of cut away = .75

Weight = $[(.7854 \times 10.375^2 \times 1.375) - (.7854 \times 7.875^2 \times .75)] \times .26$ pound

Weight = $(116.24 - 36.53) \times .26$ pound

Weight = $79.71 \times .26$ pound = *20.725 pounds*

Machining the Casting:

Accurate reproduction of the original flywheel requires the location of holes. This is a quick and easy job if the flywheel is mounted on a rotary table and a plot is made.

Centering the Original Flywheel on the Rotary Table

With an accurate plot, the casting may be machined. Center the casting in the lathe chuck using the inside surface of the ring and a surface gage. Cut the back surface flat and center drill the casting. Move the casting to the rotary table and center it using this center hole and a section of drill rod as seen in the photo below. Once the casting is properly centered, locate and drill the holes as noted in the drill plot you made from the original flywheel. Drill two additional holes for mounting the casting to the lathe chuck for finishing.

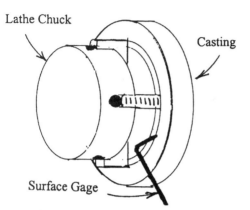

Centering the Casting on the Rotary Table

Drilling the Flywheel Casting

Finishing the Flywheel in the Lathe
Notice the two mounting molts that secure it to the faceplate.

Using the tailstock center to hold the casting against the faceplate, center and bolt the casting through the two additional bolt holes as seen in the previous photo. C – washers are helpful in securing the flywheel and are easily made by grinding through a few standard washers.

C- Washer

The finished flywheel can be balanced by setting it on shop made balancing ways. The heavy side of the wheel will roll to the bottom. Mark this location with a piece of chalk. Drill it at this location and check the balance again. When perfectly balanced, the wheel will rest in any position without rolling. (**Left:** Balancing the wheel on shop made ways.) For accurate results, the ways must be level in all directions and a ground shaft, such as drill rod must be used. Using a good level, check the ways and make a final check by setting section

of shafting between them. These ways were originally built to balance centrifugal fans for the cupola project. They work very well, but the 6-inch rulers may be a little thin for heavier wheels.

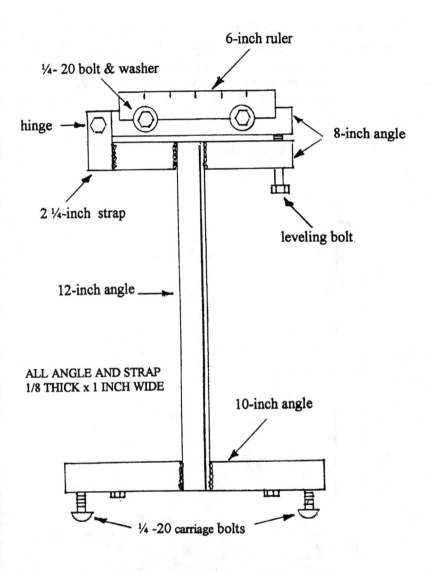

6-inch ruler

¼- 20 bolt & washer

hinge

8-inch angle

2 ¼-inch strap

leveling bolt

12-inch angle

ALL ANGLE AND STRAP
1/8 THICK x 1 INCH WIDE

10-inch angle

¼ -20 carriage bolts

Balancing Ways Side View

Set the ways up on a level surface and adjust the feet until there is no wobble. After each of the ways is leveled, check between both of the ways. If your level is not long enough to reach across the distance between the ways, use a piece of polished shaft or drill rod to span the distance. Recheck the ways until they are absolutely level.

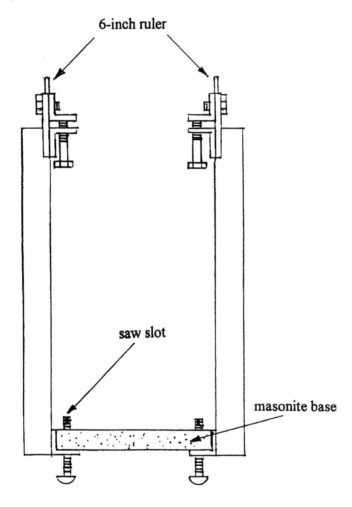

6-inch ruler

saw slot

masonite base

Balancing Ways End View

26. FLYWHEEL HOUSINGS:

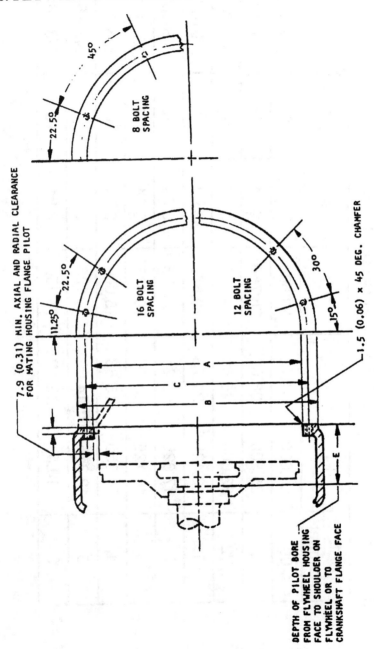

SAE Engine Flywheel Housings

SAE Number	A	B	C	E	Number of Bolts	Size
00	31.000	34.75	33.500	3.94	16	1/2-13
0	25.500	28.00	26.750	3.94	16	1/2-13
1/2	23.000	25.50	24.375	3.94	12	1/2-13
1	20.125	21.75	20.875	3.94	12	7/16-14
2	17.625	19.25	18.375	3.94	12	3/8-16
3	16.125	17.75	16.875	3.94	12	3/8-16
4	14.250	15.88	15.000	3.94	12	3/8-16
5	12.375	14.00	13.125	2.81	8	3/8-16
6	10.5	12.12	11.250	2.81	8	3/8-16

27. MAKING FLYWHEEL HOUSINGS:

Direct-coupled generators bolt directly to the bell housing or flywheel housing. Depending upon the type of generator you have, SAE or Asian, different sizes and bolt patterns are used. SAE dimensions are detailed at the beginning of the chapter. My bell housing is SAE #5, as dictated by the generator coupling.

Making bell housings, while a precision job, is not difficult. Depending upon the size of your generator coupling, your bell housing may be cast similar to an existing automotive type housing where the engine side is completely open or like the type I made with an enclosed back.

While the bell housing may be cast using thinner walls and built up bolt bosses, simplicity of the pattern and molding were the primary considerations in my design. The walls of the casting are thick enough to drill and tap, eliminating the need for a bolt flange.

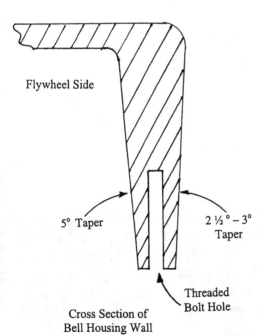

Engine Side

Flywheel Side

5° Taper

2 ½ ° – 3° Taper

Threaded Bolt Hole

Cross Section of Bell Housing Wall

A green sand core forms the large interior of the casting. Baked sand cores are used for the starter and two bolt bosses. The starter core print and core box are made from a PVC pipe coupling.

The bell housing pattern is made from yellow pine and plywood. The trick here is to get the round pattern sections to fit within the standard thickness of a 2 x 6 in pine board. Start the layout with a full sized (or properly scaled) drawing of a ¼ circular section of the bell housing casting. Using the standard thickness of the board, (accounting for a little planing to get it straight) divide the circle into sections until it fits within the available thickness.

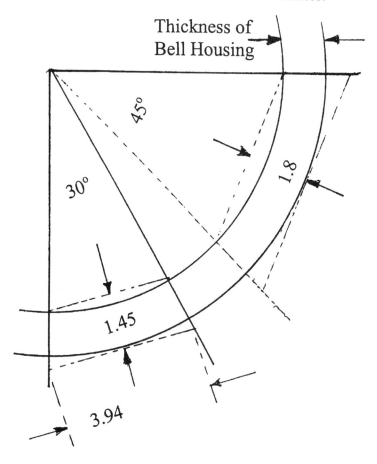

The drawing above shows that if the housing is divided into 8 sections, it will not fit within the 1.5-inch thickness of the board. However if it is divided into 12 sections, at 1.45 inches thick, it just fits. Because the ends of the board must fit together to make the 30° section, each end must have ½ of the angle. Therefore, the sections are cut 3.94 inches long and the ends are cut at 15°. My finished pattern sections are cut 4.175-inches deep to account for the thickness of the flywheel and coupling. The finished pattern dimensions at the open end are 12.45-inches ID and 14.25-inches OD. Accounting for shrinkage in the casting, the cold casting should shrink to 12.29-inches diameter, leaving a clean up cut of approximately .043-inch on the inside diameter.

> *To account for shrinkage in the aluminum casting, multiply the required finished dimensions by 1.013.* For more information on the shrinkage of castings see "Metal Casting Volume 2 P.117-118.

Cut the 12 sections required for the pattern, and one or two extra so that you can be sure to get all of your sections to fit together well. Liberally apply glue and clamp them using a band clamp. A few strips of waxed paper between the clamp and the pattern keeps the glue off of the strap. Set the glued sections on a flat surface such as the top of your table saw to dry.

Glued Sections with a Strap Clamp

Turning the Bell Housing Pattern

Glue the plywood back to the ring and secure it with a few brads or screws. Mount the pattern in the lathe and turn it to size. This pattern was over 14-inches in diameter but the capacity of my Enco lathe was only 12-inches. I had to remove the bolt-in gap in the lathe bed to fit the work on the faceplate.

There are many ways to make the cut outs in the back of the pattern and for the starter core. I mounted the pattern on a rotary table to make both the starter cutout and the large round cutout for the rear oil seal located on the crankcase.

Making the Starter Cutout

The core prints for the top two bolt-holes are measured, cut and glued into position. The starter core print is made from a 3-inch PVC pipe coupling. It is trimmed to length in the lathe and a 3 degree taper was added on a disk sander. The starter print is covered with strip of masonite and glued into position. All of the joints are coated with Bondo auto body filler and sanded for smooth fillets. The pattern is given two coats of shellac and sanded smooth.

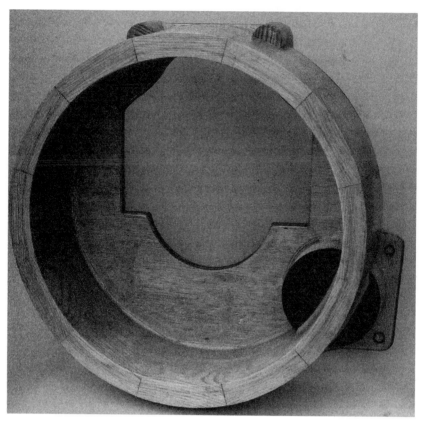

The Completed Flywheel Housing Pattern

Two core boxes are made, one for the bolt hole access and the other for the starter core. Note that the starter core has a section of PVC pipe and a strip of masonite attached to one side. This added material in the core box forms the thickness of the metal in the finished casting. See the photo on the next page and look closely at the starter core to see the two different diameters.

Bolt Cores, Starter Core and Core Box

Casting the Flywheel Housing:

The pattern is pretty large for the available flasks in my shop so I decided to locate the gates and runner system in the large cutout in the center of the casting. Because this is a large chunky part prone to shrinkage cavities, I added insulating riser sleeves around a series of blind risers. Sleeves may be purchased or made from

plaster as seen in the Metal Casting manuals. **(Left:** The risers almost completely disappear under the insulating sleeves.)** Cores were placed in their respective locations and secured with a few nails. Aluminum scrap (probably 60% piston and 40% extrusion) was melted in the tilting furnace and the casting was poured at a low temperature of 1250° F. For whatever reason, good or lucky, the first casting was perfect. The gating system was removed with a circular saw fitted with a carbide woodcutting blade.

The new casting in seen in the wheelbarrow a few hours after it was poured. Notice the gating system and the insulated risers.

A few minutes with a wire wheel, sander and file clean the casting up and have it looking good.

Machining the Casting:

Proper machining of the casting is entirely dependent upon location of the holes in the original bell housing or crankcase. In my case, the engine has already been assembled so I used the bell housing. It probably would have been easier to work from the crankcase, remembering that what you are seeing is the mirror image of what is actually needed and that the holes would be reversed on the engine side of the casting.

The original bell housing is bolted to the mill table and centered under the spindle relative to the center hole in the casting. From there, an x-y plot is made using the readout on the mill. A freehand drawing is made, recording the x-y location of each hole. At one point, there is not enough table travel to locate all of the holes. This problem is solved by mounting the bell housing on the rotary table. As many holes as possible were located and recorded, then the table was rotated 90° and the remaining holes were located.

Centering the Casting

The casting is clamped to the mill table and the rear surface (engine side) is milled flat. The casting is then moved to a rotary table and centered relative to the center of the casting and the top

Milling the Mating Surface

edge. The generator mating surface is milled flat and the inside diameter is milled using the rotary table. The critical dowel holes are located and drilled. The generator mounting holes are drilled and tapped. From here, the casting is turned over and the remaining holes are located and drilled relative to the dowel holes. Finally the starter hole is centered over the rotary table and cut to exact diameter. The casting is moved to the engine and aligned. A few whacks with a rubber mallet and it slips perfectly into position.

Cutting the Inside Diameter

Cutting the Starter Hole.
Note that the hole is centered over the rotary table.

The Completed Housing

All the holes line up perfectly and the starter properly engages the flywheel teeth. The long pin at the bottom of the starter is a temporary dowel. It will be shortened and drilled for a bolt.

28. MAKING A REMOTE STARTER:

While a key start or pushbutton starter is the easiest to build, some situations warrant a remote starter. Starting a remote set in bad weather, tying the genset starter to a air conditioning thermostat, or automatic starting by an inverter-charger system are examples where remote starting is desirable. A simple relay type starter is detailed here.

Engine protection devices such as low oil pressure and high temperature are tied to the remote starting system so that if a fault occurs, the ignition and cranking systems are turned off. This is accomplished by adding a shorting relay to a system circuit breaker.

AUTOMATIC CONTROL:

Three 12-volt, double pole double throw relays control the engine management. A small 12-volt transformer, rectifier and a capacitor power the **cranking control relay** from the generator output. The capacitor is used to keep the relay from dropping out when heavy loads are applied to the genset, such a starting a large electric motor or air-conditioner. I noticed that when the generator was loaded, starting my 4-ton AC would cause the starter contacts to bounce due to the momentary voltage drop. Addition of the capacitor solved this problem. Another bouncing problem was noted due to the residual voltage built up in the starter as the cranking circuit was opened. This voltage would arc across the relay contacts causing the starter solenoid to bounce. A 1000mf 35-volt electrolytic capacitor relived the arcing problem. While many capacitors may solve this problem, this particular value happened to be on the bench as I was troubleshooting the circuit.

On-Off Switch: This switch controls voltage to the control relay coil. *Remote circuits tap in at this point.* Closing the contacts starts the engine.

Circuit Breaker: The circuit breaker is the heart of the engine protection circuit. A drop in oil pressure or high temperature causes the protection relay to close, causing a short across this breaker, opening the circuit. I am using a 5-amp breaker. The ignition circuit requires about 3.5 amps. By selecting a breaker and resistor combination, a time delay may be designed into the system. Thermal breakers have longer trip times than magnetic hydraulic types.

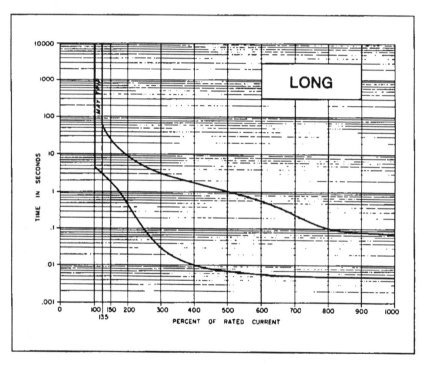

Breaker Trip Time For AIRPAX Breaker

Control Relay: A small current through this relay closes one set of contacts to energize the ignition, making it operative. Another set of contacts closes, completing the circuit to the cranking control relay contacts and engine starter or solenoid. If you are using a magneto ignition, the set of ignition contacts would be used to

152

ground the ignition in the normally closed position, turning off the engine.

Cranking Control Relay: This relay controls the engine starter and enables the engine protection devices. This relay is tied to the generator output. Cranking continues until engine speed rises enough for sufficient voltage to be built up in the generator. This activates the relay, opening the cranking contact and closing another contact that enables the engine protection circuit.

Engine Protection Relay: This relay is tied to the oil pressure switch and the temperature switch. When either of these devices closes, it actives the coil in the relay closing a set of contacts that shorts the system circuit breaker, turning off the engine.

If the oil pressure is slow to build upon starting, a manual momentary switch may be added to override this circuit and prevent the contacts from closing before the oil pressure has built up.

R1: This is a wire wound resistor to reduce the current flow through the relay contacts. Although I am currently not using this part, experimenting with the value and the breaker curve may provide a time delay for the oil pressure circuit.

Additional Equipment for Gas and Diesel Engines:

Gas engines should have an electric gas cut off valve before the regulator. While it is not essential for gensets located outside, any gas generator located indoors or in a garage *must* have a cut off to prevent gas from leaking in to the garage due to a failure of the engine or regulator. This gas valve is normally closed and requires manifold vacuum or 12-volts to hold it open. Tie the valve into the control relay ignition circuit as required.

The fuel on a diesel engine must be cut off to stop it. A fuel solenoid is used to close the fuel line. The solenoid is normally closed until 12 volts are applied, then like the gas valve, it opens. This solenoid is tied in to the ignition circuit on the control relay.

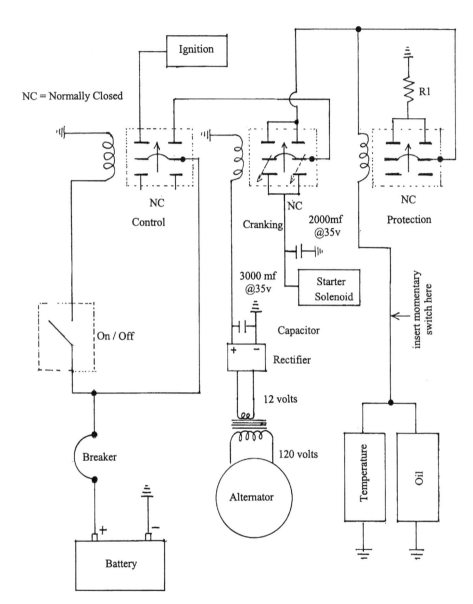

NC = Normally Closed

Ignition

NC
Control

NC
Cranking

2000mf
@35v

NC
Protection

R1

insert momentary switch here

On / Off

3000 mf
@35v

Starter
Solenoid

Capacitor

Rectifier

12 volts

120 volts

Breaker

Alternator

Temperature

Oil

+   −

Battery

Remote Starter Circuit

ALTERNATIVE WIRING FOR COIL RESISTOR:

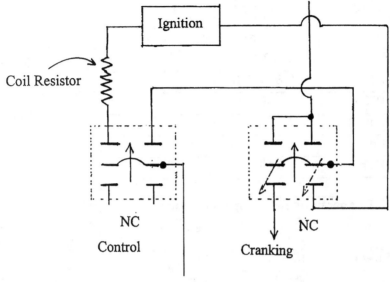

Some coils require resistors to prevent them from overheating. Such resistors are run in series with the distributor coil. Often circuits bypass the resistor during cranking to provide a hotter spark in case of low battery voltage. The addition of 1 wire to the cranking relay will bypass the resistor until the engine comes up to speed.

Coil Primary Resistance Value				
Coil Number	Voltage	Coil Resistance	Resistor number	Resistance ohms
IC 7	6	1.14 -1.26	not used	
IC 10	12	1.42 - 1.58	ICR 10, ICR 11	1.35
IC 12	12	1.25 - 1.36	ICR 13	1.82
IC 13	12	1.57 - 1.73	ICR 13	1.82
IC 14	12	3.15 - 3.48	not used	

NAPA Coil and Resistor Part Numbers

NAPA Oil Pressure Switches

Part Number	Thread	Pressure PSIG
OP6121	1/ 8 -27 NPT	1
OP6619	1/ 8 - 27NPT	2 - 7
OP6616	1/ 8 - 27 NPTF	2 - 6
OP6615	1/ 8 - GAS	.5 - 2
OP6382	1/4 - 18 NPT	2 - 7
OP6064	M12 - 1.5	1 - 2.5
OP6127	M12 - 1.5	1 - 2
OP6137	M10-1	3.75

NAPA Temperature Switch Specifications

Part Number	Thread	Temperature Closes
TS 6471	1/2 -14	248
TS 6605	3/8 - 18	255
TS 6607	1/2 -14	248
TS 6609	3/8 - 18	245
TS 6613	1/2 - 14	266
TS 6619	1/2 - 14	267
TS 6624	1/2 - 14	300
TS 6630	1/2 - 14	285
TS 6637	3/8 - 18	263
TS 6642	3/8 - 18	240

STARTER SOLENOIDS AND KEY SWITCHES:

If your engine requires a starter solenoid, Ford type solenoids are inexpensive and readily available. When the starter is engaged for extended periods of time, NAPA# ST80 is suggested. For continuos duty 6 volt systems, NAPA# ST86. I have found good key-switches at Wal Mart.

DISTRIBUTORS: You are not limited to "the parts that come on an engine" when you have a small foundry. A 5-gallon bucket furnace allows you to cast all kinds of custom parts for your project. Changing a computer ignition to a magnetic reluctor or points type distributor is not too difficult. You can cast and machine a new distributor or use parts from an old one, casting a new body. This is the case for one of my projects. A 70's Ford 1600 distributor was $34 with no core exchange. I cast a new body, machined the shaft and used it on a computer controlled Nissan. It is easiest if the replacement distributor has the same drive tang or gear. If not, it is not too much trouble to machine a new shaft. Pay attention to the rotation of the distributor, your new one should rotate in same direction. I have made a few distributors and have been pleased with the results.

Distributor parts are available from NAPA. They have an extensive electrical parts catalog that enables you to come up with just about anything. Below are a few distributor caps and rotors used on 4 cylinder engines. Screw type caps rather than those held by springs are a little easier use because you don't have to come up with a pair of spring clamps.

NAPA Part Numbers:

English Ford 1600 '66-80 – Counter-clockwise rotation
Cap = EP 50 Rotor = EP 49 (either rotation)
Point set: Counterclockwise rotation = CS755
       Clockwise rotation = CS758
Coil = IC 10 Resistor = ICR 10 or ICR 11 Condenser = EP 30 25mf
Ford Pinto 70's 2000 engine – clockwise rotation

Nissan screw type cap: '83-88 E-13, E15 E-16 engines
Counter clockwise rotation - point type and magnetic reluctor
distributors: Cap = EP 769 Rotor = EP 773

Nissan Screw type Cap '91 and later 1600 engine Clockwise rotation. Cap = EP 769 Rotor = EP 763

Assembly Notes:

Typically, the project would be assembled in a sheet-metal box. However, I had plenty of scrap aluminum and do not particularly like sheet-metal work, therefore I sand cast an aluminum box. While I did fit every thing inside, it really should have been 1 to 2-inches larger in all directions. I used a green sand core, which requires a 5° taper on the inside. The outside has a more standard 2° taper. While the sand cast box is nice, it would be less work to bend one from sheet-metal.

An inexpensive oil pressure gage and ammeter were installed in the box. I would not consider starting a generator without an oil pressure gage, *especially* a newly rebuilt engine.

The 12-volt transformer is mounted on the outside rear panel of the box. All wiring is drawn in through holes lined with rubber grommets. My ignition resistor is mounted in the inside of the box, however, because it gets warm, it might be better located on the rear panel.

There are a variety of relays available and an even wider variety of prices for the same device. I purchased 12-volt general-purpose DPDT relays and terminal strips from MSC industrial supply. They were fairly inexpensive. Nuline part # 78RCSX2, MSC Stock # 07019367. The relays were secured in the box with ½-inch wide sheet-metal strips.

29. HOMEMADE CASTINGS SOLVE PROBLEMS:

Many drive and routing problems can be easily solved with castings. Below, two problems are illustrated. First, the radiator outlet was sized differently than the stock hose.

Casting a small adapter that fits inside the bottom hose at the radiator solves this problem. The adapter pattern is seen below.

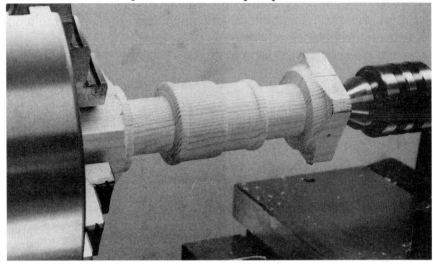

Turning the Hose Adapter Pattern

The center of the pattern converts between the two hose sizes and the extensions are the core prints for the inside diameter of the transition. The square ends are sawn off and are not part of the pattern. The outside diameters may are may not require turning after casting.

The second problem was finding a suitable drive for the governor. The crankshaft pulley has only one groove and is used to drive the water pump and alternator. Casting a two-groove water pump pulley solves the drive problem.

Typically, I cast deep pulleys with a baked sand core, however this casting may work with a greensand core and a 5° taper on the inside of the pulley. Using a green sand core reduces the amount of labor in the casting however green sand cores are weak and may break. Using a few nails to secure it usually solves this problem.

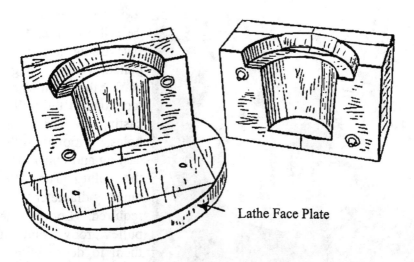

Lathe Face Plate

Casting a Deep Pulley: The pulley pattern is seen on the right. The lip "A" at the top of the pattern projects about ½ inch. The corebox seen above is turned on the lathe. Smaller core boxes may be fitted into a 4-jaw chuck. Baked sand cores work very well and this type of casting is easy to make.

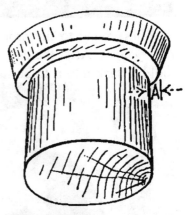

Machining the Pulley: After the casting is cool, saw off the gates and remove the sand and any fins on the pulley casting with a wire brush followed by a sander. Center the casting in a 4-jaw chuck and cut the inside face flat. Follow this by boring the inside pulley diameter as required to fit your water pump. Without removing the casting, center drill and then step drill the casting out to the water pump shaft diameter. It is best to drill the final hole 1/64th-inch small and ream the hole to size.

I made a fixture to hold the pulley in the lathe by welding a section of a junk water pump to a steel rod and then facing the

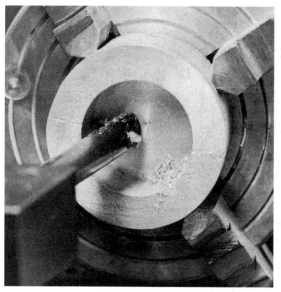

(**Left:** boring the inside of the pulley casting.) front flat. Using a transfer punch, (next page) mark the bolt hole locations in the pulley casting. Drill the holes, mount the casting and chuck it in the lathe. Turn the outer diameter as required. Locate the pulley grooves and cut them to depth with a cut off tool. Finish the belt grooves with a tool bit ground to the proper angle as seen in the SAE *Belt and Pulley* chart. After the last cut, oil the groove and lightly work the tool bit back into it to burnish the sides. Round the outer edges of

the groove with a file. (**Left:** Drilling the pulley) **Flat Single groove pulleys** do not require cores and are cast as a disc with a riser in the middle. Leave the riser attached to the casting and turn it down to mount the pulley in the lathe. Finish as required. Drill and tap the pulley for two socket head screws inside the groove. I usually use 5/16-inch screws at opposite sides of the pulley. Pulley taps are

inexpensive and long enough to tap deep pulleys. The set screws grip the shaft very tightly and make a sturdy connection. Set screws may also be used on deep pulleys, however they will be located at the front shaft, not in the grooves.

Left: Marking the bolt holes on a deep pulley using a transfer punch.

Below: Turning the outside diameter of the multiple groove deep pulley.

Finishing the Sides

Cutting Groove to Depth

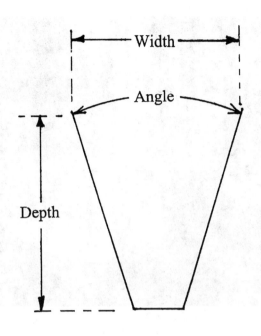

SAE Belt and Pulley Dimensions in inches

Belt and Pulley Size	Width of groove	Nominal Belt Thickeness	Pulley Diameter	Groove Angle	Minimum Depth of Groove
0.380	0.380	5/16	2.75 and up	36	7/16
0.500	0.500	13/32	3 and up	36	9/16
11/16	0.597	13/32	3 to 4 over 4 to 6 over 6	34 36 38	9/16
3/4	0.660	7/16	3 to 4 over 4 to 6 over 6	34 36 38	5/8
7/8	0.785	1/2	3.5 to 4.5 over 4.5 to 6 over 6	34 36 38	11/16
1	0.910	9/16	4 to 6 over 6 to 8 over 8	34 36 38	13/16

Using a Pulley Tap

Finishing the Sides

166

Thermostat housings: Casting a replacement thermostat housing to reroute the cooling hoses is not difficult.

The split pattern is assembled from two different sized sections turned in the lathe. The projections seen on the pattern are core prints. To keep the pattern closed during machining, a C clamp is added and the joint between the body and hose connection is drilled with a center cutting end-mill. Chuck the body of the thermostat cover in a dividing head to index the compound angle at the joint.

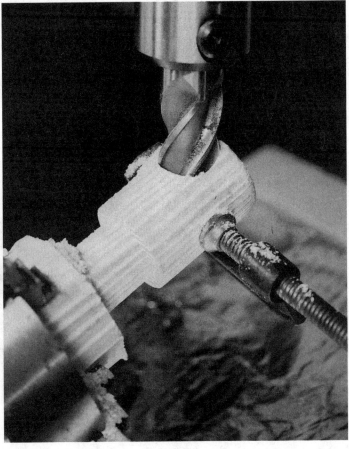

Make the corebox by drilling or boring the appropriate sized holes and smoothing the joint between them with Bondo auto body filler.

Core Box and Sand Core

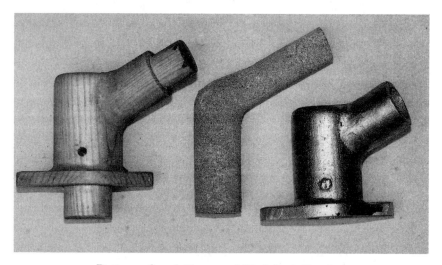

Pattern, Sand Core and Finished Casting

30. MAKING A WATER COOLED EXHAUST MANIFOLD:

The original propane generator project included a combination water-cooled exhaust manifold and header tank. A prototype manifold was cast and a few small changes were made. Because the manifold is rather heavy, a bolt boss was added to the bottom so that an additional steel support, similar to the tubing used on the governor project, could be used. All of the other changes were thickening of the bolt bosses in other sections of the heat exchanger.

The radiator cap neck on the top of the heat exchanger is a fabricated and bolt on part. A radiator neck was purchased at a repair shop and soldered to a strip of brass sheet that was left over from the shell and tube project. Later, I found that bolt on necks were fairly common and could be picked up for nothing at an auto scrap yard. There were several models of Nissans with such necks. They are also common in boating supply houses for use on heat exchangers and header tanks.

Because the manifold is water-cooled, it is made from aluminum. Although I did have a pressure tight casting, I was prepared to seal it with sodium silicate "Block Sealer" available at NAPA auto parts stores. To seal the manifold, mix the sealer with nearly boiling water and fill the header tank. Close the tank, pressurize to 15 – 20 psi and shake the tank distribute the sealer to all sides. Roll the tank every few minutes to be sure all surfaces are sealed.

Thinking I would be able to reduce the time spent on pattern making, I chose to make the exhaust manifold bolt flanges as pasted cores. While the system worked. I do not think any time was saved because of the time spent making and fitting the cores and the possibility of setting one backwards, which I did ruining an otherwise perfect casting.

While the water-cooled manifold was a success, I don't know that the time spent on it was worth the estimated 3 to 5% additional heat recovered. I have not yet made another manifold for the diesel

169

engine but instead fabricated a header tank from 4-inch pipe. A few of the manifold casting details are seen below:

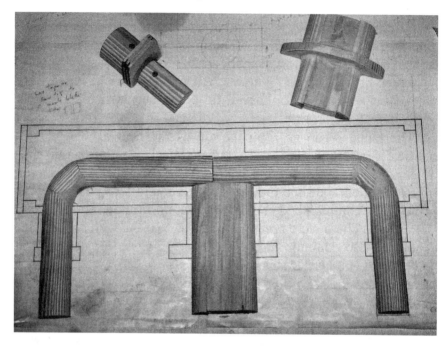

The pattern is laid out on a full sized drawing with shrinkage allowance. The top sections are mounted in small core boxes and form the mounting flanges. The bottom section is used to form a negative pattern for a plaster mold. The actual core box for this section is made from plaster. The wood pattern is discarded after making the plaster mold. Below: Core box for water side of above.

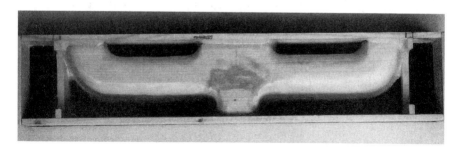

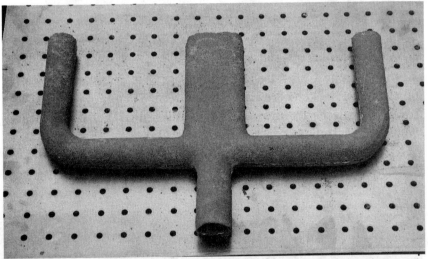

Two halves of the exhaust gas side core are pasted together and baked on a dryer plate.

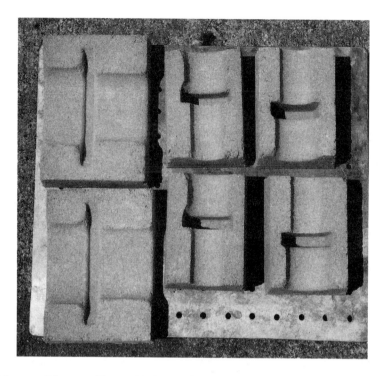

Above: Flange Cores **Below**: Pasted up Water Jacket cores. The upper water jacket cores are made from a singe core box with reversible pieces to form right and left sides. The lower section, water-side core box is seen in an earlier photo.

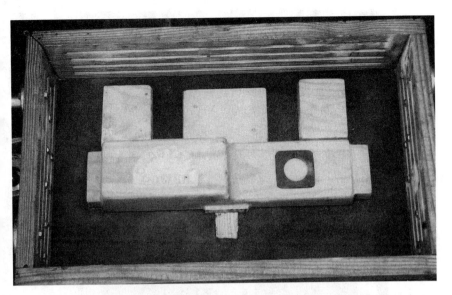

Pattern in sand mold. Drag is rammed. Notice the large core prints.

Assembling the Cores in the mold. Molding sand is rammed around the core print at the bottom of the photo to prevent floating.

Above: Completed Core assembly
Below: Machining the Casting – cutting the ends flat in the lathe.

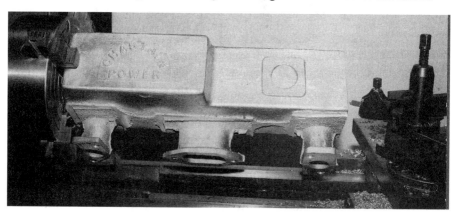

174

Section throught the finished casting. The jacket-headertank core is a little off center however there is still sufficient wall thickness so that the casting would still work.

The tabs protruding from the sides are small baffles to slow the water flow and allow air bubbles to separate. The end plates for the casting are cut from ¼-inch aluminun plate and the hose bibs are attached on these end plates.

This is a prototype casting to develop the process. Because of the weight of the casting and additional hoses and accessories, a heavy bolt boss was added to the bottom for a strut to relieve the strain in the flange bolts.

31. BASIC DEFINITIONS:

Before we start we need a few definitions. Although much of the world uses the SI units, Americans think in the British system, i.e. mile, foot, inch, pound, Btu and temperature in degrees Fahrenheit. Therefore, all the material that is presented here is based on those "familiar" Standard units.

Absolute Zero. For many calculations to properly work out, they must be based on the temperature above absolute zero. Absolute zero on the Fahrenheit scale is – 459.7°. The Rankin scale starts at absolute zero. To convert from Fahrenheit to Rankin, add 460° to the Fahrenheit temperature.

Calculate the Rankin temperature if the Fahrenheit temperature is 71°:

$$71° + 460° = 531°R$$

atm: Atmosphere. A unit of 14.7 psia, which is the pressure of the atmosphere on a normal day at sea level.

The **Btu** or British thermal unit, is the amount of heat required to raise 1 pound of water 1 degree Fahrenheit.

CFH: Cubic feet per hour.

CFM: Cubic feet per minute

Cp: Abbreviation for Specific Heat (see definition)

Flash point: As the temperature of fuel oil increases, vapor is given off and collects at the surface. When the temperature rises to a point where the vapors ignite when exposed to an open flame, it has reached its flash point. The flash point is the highest temperature at which the oil can be stored without being explosively dangerous.

The flash points of common oils are listed below:

Gasoline	Less than 100° F
Kerosene	132° F
Diesel	175 to 220° F
Fuel oil	170 to 280° F
Lubricating oils	400 + ° F

Note that some MSDS sheets list #2 Diesel at 128°F

Fire point. The temperature at which an oil gives off enough vapor to support sustained combustion is called the fire point. The fire point is usually about 50 degrees higher than the flash point.

Foot^2: Square foot (foot x foot).

Foot^3: Cubic foot (foot x foot x foot).

Heat of Fusion is the amount of heat required to melt a substance after it reaches its melting temperature. A familiar example is the melting of ice. When ice melts, heat energy is absorbed yet the temperature remains at 32° F until all the ice is melted.

There is no change in the temperature until the change of state (solid to liquid or liquid to gas) is complete.

The heat of fusion of ice is 144 Btu per pound.

The **Heat of Vaporization** is the amount of heat required to change a liquid to a vapor (boil) without a change in temperature. The heat of vaporization of water is 970 Btu per pound. The heat of vaporization of propane is 183 Btu per pound, Gasoline-126 Btu per pound and Diesel – 140 Btu per pound.

Laminar flow occurs when the flow is smooth and travels in layers that do not mix or swirl. Smoke rising from a cigarette in still air resembles laminar flow.

177

psia: The pressure measured relative to an absolute vacuum

psig: The pressure relative to the atmosphere, or pressure above atmospheric pressure. Most pressure gages read zero at atmospheric pressure

Rankin: Temperature above absolute zero. Add 460 to the Fahrenheit temperature to find the ranking temperature.

Specific gravity: The weight of a substance relative to an equal volume of water. Lead has a specific gravity of 11.4. If a container holds 1 pound of water, it would hold 11.4 pounds of lead.

Specific heat is the number of Btu's required to raise the temperature of 1 pound of a substance 1 degree Fahrenheit. The specific heat of a substance changes at high temperatures and with changes of state.

Specific Heats of Common Substances at Room Temperature

Water	1.000
Air	.240
Iron	.110
Aluminum	.248
Copper	.093

Calculate the number of Btu required to raise the temperature of 5 pounds of iron from 60° to 70° F:

Number of degrees = 70 – 60 = 10

Number of Btu = weight x specific heat x number of degrees

Number of Btu = 5 pounds x .110 Btu / pound x 10

Number of Btu = 5.5

This calculation may be carried out for any substance with a known specific heat.

Standard Cubic Foot: 1 cubic foot of gas at standard temperature and pressure, 77°F and 14.7 psia.

The **State** of matter refers to the condition it is in i.e. solid, liquid, or gas.

Thermal Conductivity is the ability of an object to conduct heat. If a copper rod is held in one hand and the opposite end is heated, soon you will feel the temperature rise. If a wooden stick is heated in the same manner, you will not feel any change in temperature. Copper has a high thermal conductivity and wood has a low thermal conductivity.

Turbulent flow occurs when there is random motion of the flow, swirling and mixing.

Wetted Area or Surface: The area inside a tank covered by liquid.

32. VOLUMETRIC EFFICIENCY:

In real life, engines do not achieve their "theoretical performance." One of the reasons why a four-stroke engine does not achieve an ideal cycle is that *volumetric efficiency* is usually between 80% and 90%. Volumetric efficiency is the amount of air fuel mixture *actually* entering an engine relative to the amount that would enter the engine under *ideal conditions*.

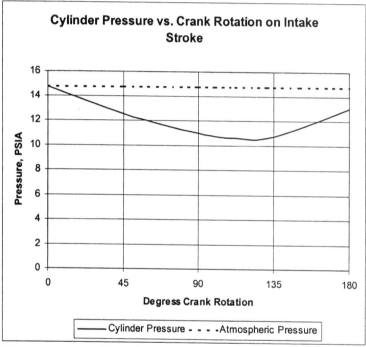

There are a number of reasons why an engine receives less than a full charge of air-fuel mixture. Because the air moves through a hot engine, it expands, therefore less air is required to fill the cylinder. Another reason is that the intake stroke lasts only about .017 second at 1800 rpm. Therefore the pressure in the cylinder is always less than atmospheric at the end of the intake stroke. The higher the engine's speed, the less time available for the air to enter the cylinder. This limits the maximum speed of an engine.

180

The engine reaches a point where the volumetric efficiency is so low that the engine can produce only enough power to overcome the internal losses (friction). This is the engine's top speed.

Example: Assume that a 24.4 cubic inch cylinder holds .0169 ounces of air at atmospheric pressure. At 1800 rpm only about .0139 ounce can enter. What is the volumetric efficiency?

Volumetric efficiency = .0139oz / .0169oz = 82.2%

33. PRINCIPLES OF COMBUSTION:

Combustion, or burning, is the rapid combination of fuel and oxygen that releases heat. Actual combustion reactions are complex, however simple combustion equations serve well for analysis of the reaction. We will be burning gasoline, fuel oil or propane, all of which are hydrocarbons so the analysis is similar. Fuel oil may contain small amounts of sulfur, however it contributes so little to the final product it is ignored in our calculations.

When carbon is burned with oxygen the reaction is:

$C + O_2 \rightarrow CO_2 + 14,093$ Btu per pound of C

When Hydrogen is burned with oxygen the reaction is:

$2H_2 + O_2 \rightarrow 2H_2O + 122,000$ Btu per 2 pounds of H_2

The combustion equations allow us to predict the amount of air required, the amount of heat generated by the combustion of a fuel, and the weights of the products of combustion.

Elements and Compounds found in Combustion

Substance	Symbol & Weight		Molecular Symbol & Weight	
Hydrogen	H	1	H_2*	2
Carbon	C	12		
Oxygen	O	16	O_2*	32
Nitrogen	N	14	N_2*	28
Water Vapor			H_2O	18
Carbon Dioxide			CO_2	44
Carbon Monoxide			CO	28
Propane			C_3H_8	44
Octane**			C_8H_{18}	114

* gases often exist as two atoms attached to form a molecule
**octane is often used to estimate liquid fuels

Calculate the reaction for burning propane, find the required amount of air in pounds and cubic feet at sea level[1] and find the gross amount of heat liberated:

$$C_3H_8 + 5O_2 \rightarrow 3CO_2 + 4H_2O$$

(36lbs C & 8 lbs. H_2) + (160 lbs. O_2) $\rightarrow$ 132 lbs. CO_2 + 72lbs H_2O

Carbon = 3 x 12* = 36   *these values are taken from the table above
Hydrogen = 8 x 1* = 8
Oxygen = 5 x 32* = 160

If the above equation is divided by 44, the molecular weight of propane from the table on the preceding page, we find:

1 pound of propane requires 3.63 pounds of oxygen. Combustion produces 3 pounds of carbon dioxide and 1.63 pounds of water.

Air is 21% oxygen and 79% nitrogen by volume. For every cubic foot of oxygen there are 3.76 cubic feet of nitrogen. By weight, for every pound of oxygen there are 3.31 pounds of nitrogen.

Calculating the weight of air for the propane reaction:

3.31 x 3.63 lbs. = 12.02 lbs. Nitrogen

The weight of air = 3.63 lbs. Oxygen + 12.02 lbs. Nitrogen = 15.65 lbs.

For every pound of propane you need 15.65 pounds of air.

To find the cubic feet of air, divide the weight of air by the density of air at sea level:

The density of air at sea level = .0763 pounds / cubic foot

15.65 pounds / .0763 pounds/ $foot^3$ = 205 $foot^3$

For every pound of propane you need 205 cubic feet of air.

For every cubic foot of propane you need 23.82 cubic feet of air
(at sea level and 60° F)

1 lb. of propane equals approximately 8.36 ft^3 gas

1 gallon of propane equals 36 ft^3 gas

The density of air is lower at higher altitudes check the table of altitudes and densities to calculate for different altitudes.

Variation of Air Density With Elevation

Elevation above Sea Level in Feet	Air density in pounds / cubic foot at 60 deg. F	Barometric Pressure in psia	Barometirc Pressure in Inches Hg
0	0.0763	14.70	29.92
500	0.0749	14.42	29.40
1000	0.0734	14.13	28.83
1500	0.0721	13.88	28.31
2000	0.0710	13.67	27.82
2500	0.0695	13.38	27.28
3000	0.0683	13.15	26.81
3500	0.0671	12.91	26.32
4000	0.0659	12.69	25.84
4500	0.0645	12.42	25.35
5000	0.0635	12.21	24.89
5500	0.0622	11.98	24.41
6000	0.0612	11.79	23.98
6500	0.0600	11.53	23.51
7000	0.0588	11.32	23.08
7500	0.0578	11.12	22.63

The approximate gross amount of heat released per pound of propane is:

For the carbon reaction:

36 pounds x 14,093 Btu / 44 pounds = 11,530 Btu

For the hydrogen reaction:

4 pounds x 122,000 Btu / 44 pounds = 11,090 Btu

The gross heat released by burning one pound of propane is:

11,530 Btu + 11,090 Btu = 22,620 Btu Gross heating value

Nitrogen functions as "dead weight" in the reaction, absorbing heat from the combustion process thereby lowering the flame temperature.

We have calculated the theoretical gross heating value or higher heating value. This value might be attained in a lab under controlled conditions, starting and ending the experiment at $59°$ F, however it is unlikely that one would ever get this in actual practice, especially since the exhaust gasses are commonly $1200°$ F or higher. Water, created as steam in high temperature combustion has stored in it the "heat of vaporization" which is the energy required to turn the liquid into vapor. Each pound of steam absorbs about 970 Btu. When the water in the exhaust leaves as vapor, all the available heat has not been extracted. This is called the lower heating value. When the water leaves as liquid, all the heat of combustion has been removed and this is called the higher heating value.

Combustion losses are due to less than ideal mixing of the fuel and air and the cooler walls of the combustion chamber quenching the flame among other things. Less than the ideal amount of air results in incomplete combustion. With decreasing amounts of air, increasing amounts of carbon monoxide are formed. Carbon monoxide only releases 4,347 Btu per pound of carbon burned as opposed to 14,093 if carbon dioxide were formed. In the real world, stoichiometric, or perfectly proportioned mixtures, all of the fuel molecules will not find all of the oxygen molecules for "perfect combustion."

Elevation of Cities over 1000 feet above Sea Level	
Albuquerque, NM	4950
Atlanta, GA	1050
Canton, OH	1030
Denver, CO	5280
El Paso, TX	3695
Oklahoma City, OK	1195
Omaha, NE	1040
Phoenix, AZ	1090
Salt Lake City, UT	4390
Santa Fe, NM	6950
Spokane, WA	1890
Tucson, AZ	2390
Wichita, KS	1290
Mexico City, Mexico	7349

34. AN OUTLINE OF THE HEAT EXCHANGER DESIGN AND CONSTRUCTION PROJECT:

Evaluation of the factors considered in heat exchanger design may appear to ramble on forever with no apparent connection to the end result, a *working generator set.* In order to provide a sense of direction, the following outline is provided:

1) Determine the horsepower required to generate 15kW.

2) Estimate the efficiency of the engine.

3) Determine the amount of heat, in Btu's, required by the engine.

4) Using the Btu content, determine the amounts of various fuels required by the engine.

5) Using the volumetric efficiency of the engine, determine if is suitable for the generator project or if it requires de-rating.

6) Estimate the load on the cooling system, the heat lost through the exhaust and the radiated heat.

7) Determine the water temperature and flow through the engine water jacket.

8) Estimate the mass flow of exhaust gas.

9) Design a water-cooled exhaust manifold and estimate the heat loss to the manifold when cooled by the engine jacket water.

10) Design an exhaust gas heat exchanger based on the exit temperature of the manifold water.

11) Design awater to water heat exchanger to transfer the recovered engine heat to the hot well or process load.

12) Build the heat exchangers.

13) Test the unit.

14) Miscellaneous considerations

One could skip right to the construction chapters, however you would not have an understanding of *why* the decisions regarding the design were made. After reading the more theoretical sections of the book, you should be able to change the size of the generator set with predictable results. This allows you the freedom to design any type of powerplant you need. As with many theories, there is a gap between the theory and what happens in real life, however the trends represented here provide a good basis for design decisions. All this means, even if the theories and formulas used here were more accurate predictors of the real life situation, "we would still make the same design decisions." Math is introduced only to the point that it helps with an explanation. The math is not difficult and may be performed on a $8.95 Ti-25 calculator (or equivalent) available at Wal-Mart. The most difficult material is reduced to tables and graphs providing quick and easy answers to calculations that are otherwise long and tedious.

35. SELECTING AN ENGINE:

Locally, scrap engines go for $49. Two engines were selected for the generator project, a 1600cc Ford Pinto engine and a Nissan 1600cc Gai16 engine. While several small engines were available, these were selected because they had chain drive camshafts as opposed to belt drives. Belts require maintenance, making them less desirable for this application.

The engines were disassembled, cleaned and inspected. The cylinders were not badly worn and required .002 inch or less honing to remove any taper and present a proper surface. Locally, honing on a Sunnen machine cost $5 per cylinder. Valves were ground in the lathe and hardened exhaust seats were installed in the Ford head for $7 each. Head-sets were found on e-bay for $24 and $40. Rings sets were $35 and $48. Adding bearings, gaskets, filters and electrical parts I estimate the final cost for each engine to be $300 to $350. Not bad for a completely rebuilt engine.

36. ENGINE ANALYSIS:

The advertised horsepower of an engine is not what is actually available for power generation. They often advertise what an engine could produce under ideal conditions. This usually does not include loads such as pumping the oil, coolant, running the radiator fan or the 12-volt alternator. To begin the analysis, we will calculate the power required to run the 15kW alternator and then perform a brief analysis of the engine to see if it worth a more detailed study.

1 hp = .746 kW

15 kW / .746 = 20.1 hp

20.1 horsepower are required before accounting for the generator efficiency.

Review of the generator technical data specifies efficiency of 83.7% at full load. This increases the required horsepower.

20.1 hp / .837 = 24.023 horse power

Assuming 2 horsepower are required to run the fan and water pump, the total required horsepower is:

24.023 hp + 2.00 hp = 26.023 hp

Approximately 26 horsepower are required to generate 15kW in this situation.

Now for a quick analysis of the Ford engine:

The Ford manual specifies 75 brake horsepower at 5000 rpm.

Assuming horsepower is proportional to the rpm:

75 hp / 5000 rpm = .015 hp / rpm

The four-pole generator requires 1800 rpm.

(1800 rpm) x (.015 hp / rpm) = 27 horsepower

The engine produces approximately 27 horsepower at 1800 rpm.

Although this is a rough approximation, it suggests that it may be possible to power the generator with this engine. Had the engine produced less than 26 horsepower, it would have not been suitable for further analysis.

37. THERMODYNAMIC ANALYSIS:

A simplified thermodynamic analysis of the engine will allow us to estimate the fuel consumption, the required combustion air, the exhaust flow and the heat generated by the engine. These values form the basis for the heat exchanger design.

Efficiency of the Ford engine is 26.4%

(The average efficiency of three Ford engines of similar compression ratio was 26.4%)

1 horsepower = 2545 Btu / hour

There are 115,000 Btu per gallon of gasoline

There are 2365 Btu per cubic foot of propane

There are 909.8 Btu per cubic foot of methane

Air is at 70°F and has a density of .075 lb/ ft^3

The required Btu per hour are:

2545 Btu / .264 (efficiency) = 9640 Btu / hp hr

(27 hp) x (9640 Btu / hp) = 260,280 Btu per hour

The engine requires 260,280 Btu per hour

The fuel input per hour can now be estimated:

Gasoline consumption per hour:

(260,280 Btu / hr) / (115,000 Btu / gallon) = 2.26 gallons / hr

The engine requires 2.26 gallons of gasoline per hour at full load.

Propane consumption per hour:

$(260{,}280 \text{ Btu / hr}) / (2365 \text{ Btu / ft}^3) = 110 \text{ ft}^3 / \text{hour}$

Propane fuel input is 110 ft³ / hour at full load.

Methane consumption per hour:

$(260{,}280 \text{ Btu / hr}) / 909.8 \text{ Btu / ft}^3 = 286.08 \text{ ft}^3 / \text{hour}$

Methane input is 286 ft³ / hour at full load

Combustion Air:

In order to judge the practicality of the solutions, a reference value should be obtained for the combustion air. The amount of air consumed by the engine under ideal conditions will be calculated and then adjusted by a reasonable volumetric efficiency. Typical volumetric efficiencies are in the range of 80% to 90% .

Find the amount of air consumed by the 4 cylinder Ford engine at 1800 rpm:

Bore = 3.188 inches   Stroke = 3.056 inches

Volume in^3 / min = rpm x (.7854 bore^2) x (stroke) x (cylinders) / 2

Volume in^3 / min = (1800) x (.7852 x 3.188^2) x (3.056) x 4 / 2

Displaced volume available for combustion air = 87,818 in³ / min

Converting to cfm: $(87{,}818 \text{ in}^3) / (1728 \text{ in}^3 / \text{ft}^3) = 50.82 \text{ ft}^3 / \text{min}$

The total volume displaced in 1 minute = 50.82 ft³

Assuming a volumetric efficiency of 82%

$.82 (50.82 \text{ ft}^3 / \text{min}) = 41.67 \text{ ft}^3 / \text{min}$

A realistic figure for the total fuel and air input is 41.67 ft³ / min

Estimation of air and fuel for gasoline:

The chemically correct mass ratio of air to gasoline is 14.88, assuming a slightly rich mixture, we will use 14 to 1.

If the engine uses 2.26 gallons of gasoline per hour at 5.86 pounds per gallon, then the weight of gasoline per minute is:

(2.26 gallons) x (5.86 pounds) / 60 minutes =.220 pounds / minute

The weight of air per minute is:

14 x .22 pounds / minute = *3.08 pounds of air per minute*

The volume of air per minute is:

3.08 pounds / .075 pound per foot^3 = 41.067 ft^3 / minute

The volume of air per minute for gasoline = 41.067 ft^3

The volume of .220 pounds of gasoline is .005 ft^3

This closely matches the estimated number on the previous page.

Estimation of combustion air for propane:

Propane consumption, as calculated earlier, is 110 ft^3 / hour

Propane consumption per minute is (110 ft^3 / hour) / (60 min/ hour)

Propane use is: 1.833 ft^3 / min

1 ft^3 propane requires 23.821 ft^3 combustion air

Combustion air is 1.833 ft^3 x 23.821 = 43.664 ft^3 / min

Adding the volume of fuel and air:

1.833 ft^3 Propane + 43.664 ft^3 Air = 45.497 ft^3 fuel and air

The total propane fuel and air input is: 45.497 ft³

While you may get this into the engine in real life, generally, engines run on propane are de-rated 7 to 10%.

Calculation of combustion air for methane:

Methane consumption, as calculated earlier, is 286 ft³ per hour.

Methane consumption per minute is:

(286 ft³ per hour) / (60 minutes / hour) = 4.767 ft³ min

1 ft³ methane requires 9.528 ft³ combustion air

Combustion air is: (4.767 ft³) x (9.528) = 45.417 ft³ / min

Adding the volume of fuel and air:

4.767 ft³ methane + 45.417 ft³ air = 50.184 ft³ / min

The total volume of methane and air is: 50.184 ft³ / min

The unit will surely have to be de-rated when operated on methane.

THE ENGINE HEAT BALANCE:

While each engine design is different, generally the heat flow in a typical engine breaks down as follows:

	Spark Ignition:	Diesel:
Brake horse power:	26%	36%
Cooling system:	30%	27%
Radiation:	7%	7%
Exhaust:	37%	30%

Note that at part loads a greater portion of the fuel heating value is absorbed into the coolant. *Typically, at light loads, the coolant heat transfer rate is 2 to 3 times the brake horsepower output.*

EXHAUST TEMPERATURE:

Exhaust gas temperature varies with the compression ratio and load among other things. Higher compression ratios allow greater expansion of the combustion gases during the power stroke; therefore more heat is converted to mechanical power. Generally, the exhaust gas temperature is lower in engines with higher compression ratios.

Exhaust Gas Temperature and Compression Ratio*		
Compression Ratio	Exhaust Gas Temperature Degrees F	
5 to 1	1350	air cooled
6.5 to 1	1325	
7 to 1	1350	air cooled
8 to 1	1300	
8 to 1	1200	
9 to 1	1200	
9 to 1	1100	
15 to 1	805	turbo
16 to 1	1015	
17.8 to 1	1094	
19 to 1	800	
23 to 1	735	
*Published data from several different makes and models of small gasoline and diesel engines.		

195

The Ford engine has a compression ratio of 8 to 1. Assuming that the exhaust gas temperature is similar to that in the previous chart at an 8 to 1 compression ratio we will conservatively use 1215 °F for the heat exchanger calculations.

An interesting note, I found this Ford power curve for a 1600cc engine about two years after completing my engine conversion. I was happy to see how close our analysis comes to the published data.

Liquefied Petroleum Gas

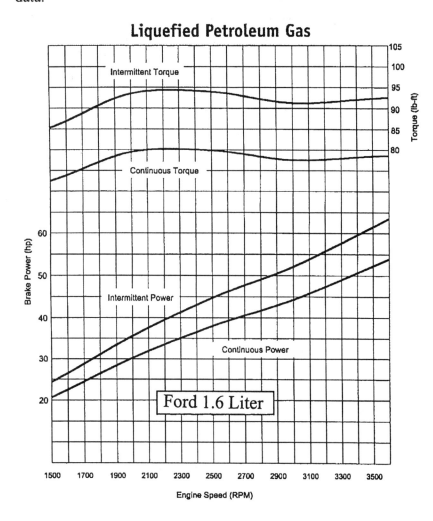

38. PROPERTIES OF FUELS:

Fuel	Specific gravity	Weight per gallon pounds	Btu/pound heating value		Btu /gallon
			higher	lower	lower
Ethyl Alcohol	.790	6.59	12,820	11,580	76,317
Gasoline	.702	5.86	20,460	19,020	111,457
Gasoline	.739	6.16	20,260	18,900	116,424
Kerosene	.825	6.88	19,750	18,510	127,349
Light Diesel	.876	7.30	19,240	18,250	133,225
Medium Diesel	.920	7.67	19,110	18,000	138,060

Properties of Fuel Gases: 36 ft ^{3} gas per gallon of Propane

Name	Formula	lb/gallon @ 60° F	ft³/ lb	lb/ft³	Btu / ft³ lower	Btu / lb lower	Btu / ft³ (air + gas)	Air ft³ /ft³ gas	Sp. Gravity air =1
Methane	CH₄		23.56	.0424	909.8	21,433	86.4	9.528	.554
Propane	C₃H₈	4.24	8.36	.1196	2365	19,770	95.3	23.821	1.562

Octane Rating and Lower Heating Value Btu/ft^3 of Gaseous Fuels:

Natural gas (Methane)	115 – 120	910
HD-5 Propane	95	2365
Butane	80	3093
Digester Gas (sewage gas)	115 – 120	600
Land Fill Gas	115-120	425

Maximum Compression Ratios for Gaseous Fuels:

Methane	15 to 1
Propane	12 to 1
Butane	6.4 to 1

39. COOLING SYSTEMS:

Inside the engine, the temperature of combustion may reach 4500°F and approximately 1/3 of the heating value of the fuel is absorbed by the engine components, which must be removed by the cooling system. The purpose of the cooling system is to keep the engine at its most efficient operating temperature under varying loads The carbonization temperature of the lubricating oil and the softening point of aluminum determines the engine's upper temperature limit. HD oils allow the top ring groove of a piston to run at 400°F and intermittently to 500°F at full power. Aluminum has good low temperature strength but looses about 50% of its strength above 600°F. Aluminum's abrasion resistance is also low at high temperatures.

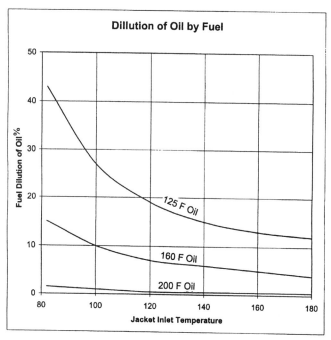

Operating an engine at too low a temperature lowers the efficiency, reduces the fuel economy, increases oil dilution and the formation of sludge and rust Low operating temperatures speed piston ring and cylinder wall wear and increases pollution by quenching the combustion next to cold combustion chamber surfaces. If an engine is cold, fuel will not evaporate readily and some of it condenses on the cylinder walls, eventually working its way past the piston rings and into the crankcase where it dilutes the engine oil. For every gallon of fuel burned, about a gallon of water is produced as steam. If the engine is hot, the steam passes out of the engine in the exhaust. If the engine is cold, some of it will condense in the engine mixing with the oil, forming sludge. Water and fuel are removed from the engine by the crankcase ventilating system after the engine reaches operating temperature. It is important for the engine to reach operating temperature as quickly as possible.

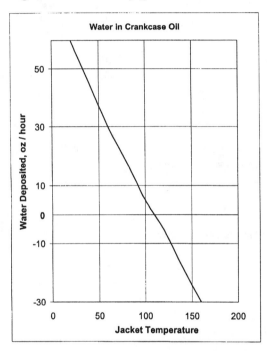

The engine cylinders and heads are surrounded by water filled jackets. The heat removed by the cooling system first flows through a very thin layer of stagnant gas adjacent to the metal walls inside the combustion chamber. It then flows through the metal walls to the water. Each step offers resistance to the heat flow. The stagnant layer of gas has more resistance to the flow of heat than all the other parts of the cooling system combined. The water pump circulates the water through the engine and cooling

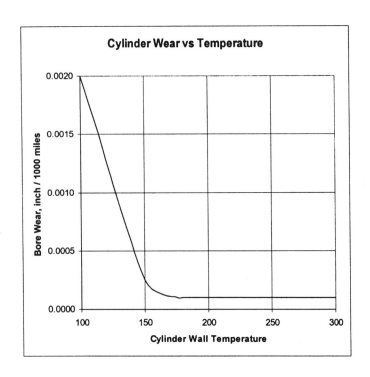

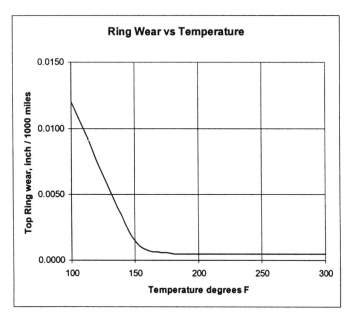

system. Heat is finally removed by a heat exchanger that may be either a shell and tube type or a common water to air radiator.

There are two basic types of radiators: the down flow and the cross flow. In down flow radiators, the coolant flows from the top of the radiator to the bottom. This radiator has a tank at the top and one at the bottom. The cross flow radiator has the supply tank on one side and the collecting tank on the other. The flow is from side to side. This type of radiator is much shorter than the down flow type, making it popular in automotive applications.

DE-AERATION OR AIR SEPARATION: It is important for the cooling system to eliminate any air that might be circulating with the coolant. The thermostat usually has a small hole to bleed the air trapped in the engine's water jackets after filling. (Many cylinder

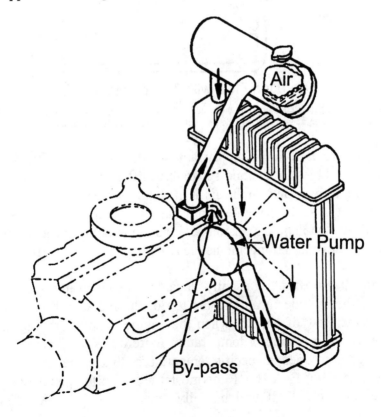

heads have an air bleed screw located at the highest point in the water jacket. The air should be bled from this point during initial filling or flushing of the coolant system.)

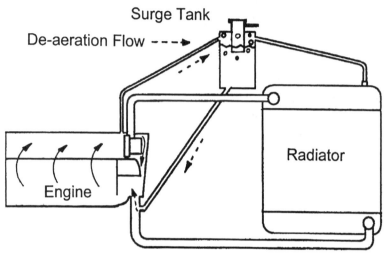

De-aeration Flow with Thermostat Closed

Note that this surge tank is configured for the bleed line before the thermostat. If the bleed line were located on the water pump by-pass circuit, then it would enter the base of the surge tank.

Baffles are used in the supply tanks to slow the flow of coolant and allow bubbles to rise and separate. Baffles also prevent water from being thrown up into the air at the top of the tank.

SURGE TANKS or separate supply tanks: Additional volume is required for coolant expansion and air separation (deaeration). There are several arrangements depending upon the thermostat design.

The **thermostat** is usually located in the water passage between the cylinder head and the top of the radiator. It stops or reduces the flow of coolant when the engine is cold, causing the engine to reach operating temperature more quickly. Thermostats are designed to open at specific temperatures that range from 160° F, 180° F or 190° F. A 160° F thermostat will start to open between 157° F and 162° F. It will be fully open at 182° F. A 180°F thermostat opens between 177° F and 182° F and is fully open at 202° F. Thermostats are selected to suit both the engine and antifreeze requirements. With glycol type antifreeze, a 180° F thermostat is used while alcohol based antifreeze uses a 160° F thermostat.

PRESSURE CAP: At sea level, where the atmospheric pressure is 14.7psia, water boils at 212° F. At higher altitudes, the pressure is lower and water boils at lower temperatures. Higher pressures cause the water to boil at higher temperatures. Each pound increase in pressure increases the boiling point by approximately 3¼° F. An increase in the boiling point allows the coolant to run at a higher temperature. The process is more efficient because heat transfer to the air increases with a greater difference in temperature between the coolant and the air. Pressure caps may provide as much as 15 psi increase in pressure and will increase the boiling point to 250°

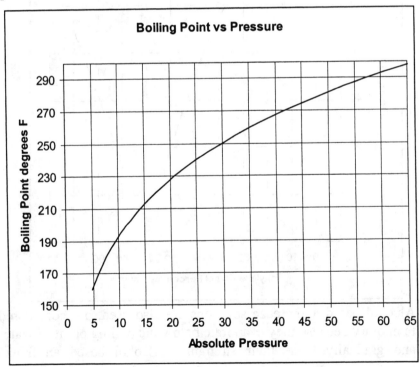

F. A 15-psi cap may increase the cooling system efficiency as much as 30%.

ANTI FREEZE: Water freezing in the radiator and water jackets would burst the radiator and crack the engine jackets. Anti freeze is used to prevent such freezing and engine damage. The most common anti freeze is ethylene glycol.

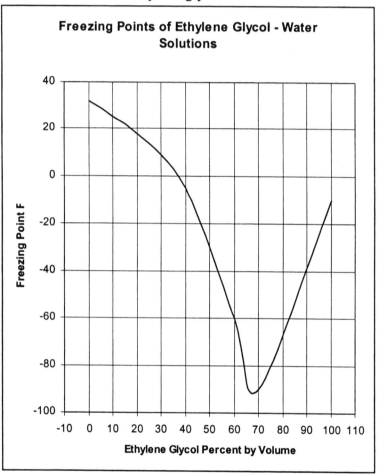

Alcohol based anti freeze solutions are also used, but considered temporary because they evaporate below the boiling point of water and gradually loose their strength.   Alcohol based antifreeze requires periodic additions to maintain adequate freeze protection. Antifreeze is mixed with water in various proportions depending upon the lowest expected temperature. With lower the temperatures requiring a higher percentage of antifreeze. *Antifreeze solutions have a lower rate of heat transfer than pure water* as will be seen in the heat exchanger design chapter.

Anti freeze should be changed at least every other year because it deteriorates overtime causing electrolytic corrosion of the cooling system. Using a digital voltage meter, check for potential voltage by inserting one end of a test probe into the coolant and grounding the other on the engine. Magnesium electrodes may be used to provide cathodic protection of the engine and cooling system components. *Diesel Rated Coolant* is specially formulated to prevent pitting of cylinder liners. Soluble oil is NOT recommended as a corrosion inhibitor. A 1.5% solution of soluble oil increases the fire deck temperature by 6% and a 2.5% solution increases the temperature by 15%.

COOLING SYSTEM SIZE: Cooling systems are designed with excess capacity so that they will function even though the radiator or water jackets have become partially clogged. However, the water jackets are not made excessively large so that a large amount of water will not have to be heated before the engine is warmed up. Typically, at least 10% excess capacity is added for system deterioration over time.

SYSTEM COOLANT CAPACITY: The total system coolant capacity must be known in order to determine the expansion and deaeration volume. The coolant capacity includes the engine jackets, the radiator or heat exchangers and the associated plumbing. A minimum of 6% expansion volume must be provided in the top tank and least 2% deaeration volume.

AIR INTAKE AREA REQUIREMENTS: When the generator is loaded, the air temperature leaving a radiator is between 150 and 185° F. When combustion air intake temperature exceeds 120° F many engines have a significant loss of power, therefore engine room or enclosure ventilation is important. If the generator set has an engine mounted radiator, typically the air intake of the

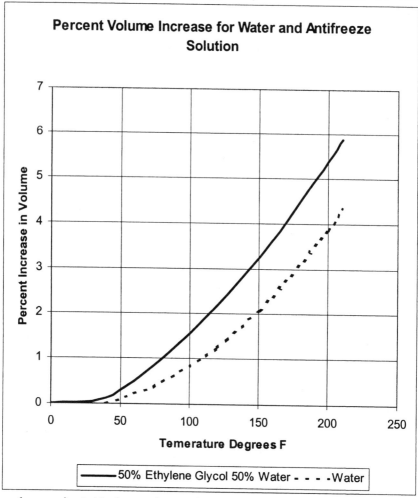

Percent Volume Increase for Water and Antifreeze Solution

Y-axis: Percent Increase in Volume
X-axis: Temerature Degrees F

Legend: ——50% Ethylene Glycol 50% Water - - - -Water

enclosure is 1 ½ times the radiator core area. If the radiator fan delivery is known, then the maximum velocity at the intake should be less than 30 feet per second or 20 miles per hour.

ROOM VENTILATION REQUIREMENTS FOR REMOTE RADIATORS: Radiators may be mounted away from the engine and driven by a thermostat controlled electric fan, however the engine room still requires ventilation. The heat produced by the generator head and radiated by the engine at rated load is approximately 35% of the kW rating of the set. Rooms or enclosures without a radiator fan, require auxiliary ventilation to remove the excess heat from the room. The required airflow is typically based on an ambient, or intake temperature of 105° F and a maximum temperature rise of 15° F. The required air flow at rated output at 500 feet above sea level may be estimated by:

Minimum Required Air flow, $CFM = 1000 \times kW / 10.75$

Estimate the required airflow for a 15 kW gas generator set operating at rated output and located in a garage using a remote radiator (assuming 7% radiated engine heat and generator efficiency of 83%):

15 x 1000 / 10.75 = 1395 CFM

The required airflow =1395 CFM

After checking the Enco Tools catalog, a 12-inch 1/12-hp shutter type exhaust fan easily meets the required airflow to limit the rise in room temperature at or below 15° F. (USE-ENCO.COM)

If there is a choice, a generator set mounted radiator is preferred.

There are two types of remote radiators: Radiators located horizontally from the engine and those located vertically from the engine. The primary concern with horizontally located radiators is that the piping system's resistance to flow not exceed the engine's water pump capacity. Pump capacity may be calculated or is usually available from the engine manufacturer. Piping should be larger in diameter than the water pump inlet diameter and as few bends and couplings as possible. Flexible couplings (hoses) should be used between the engine and the piping system to allow movement and thermal expansion of the system.

207

The most common remote radiators are located above the engine with the height being limited by the pressure of the water on the hoses and seals and gaskets in the cooling system.

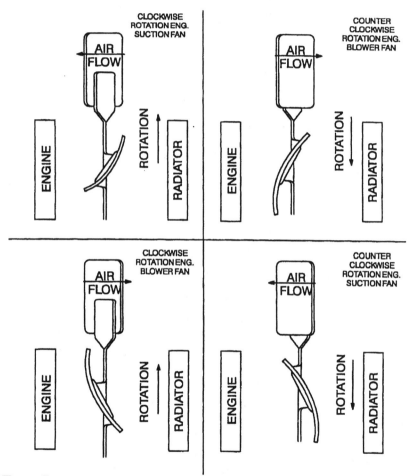

FAN SELECTION: Using the largest fan diameter possible, operating at a lower fan speed is more economical and produces less noise, however the fan blades should never extend beyond the radiator core. Blower fans are generally more efficient relative to power expended per unit mass airflow. This is because the air

208

entering a suction fan is heated by the radiator to 150 to 185° F, while the air entering an engine mounted blower fan is closer to ambient temperature. Generally, suction fans will have the concave side of the blade facing the engine while a blower fan will have it facing the radiator. A suction fan can not be converted into a blower fan by reversing the blades. The fan rotation must also be correct. Because often the water pump and fan are mounted on the same pulley, there are limits to how much one may change the fan speed relative to the original fan design.

FAN DISTANCE: Maintain at least 2 to 4-inches clearance between the fan and the radiator core. If the fan area is much less than the radiator core area, then the fan should be positioned farther away to allow the airflow to spread over the entire core area.

FAN SHROUD: Fan shrouds greatly improve the cooling system performance. A good fan shroud distributes the air across the radiator core more uniformly and prevents recirculation of hot air. *Suction fans blades should extend 2/3^{rds}* to full width of the blade thickness into the shroud, while *blower fans should extend 1/3rd* to ½ into the shroud.

ESTIMATING COOLANT FLOW: The required coolant flow to remove the heat from the engine's water jacket depends upon the horsepower and allowable coolant temperature rise of the engine. A value of 10° degrees F. temperature rise is common. Coolant flow rate may be calculated from a thermodynamic analysis of the engine or estimated from the horsepower. Generally, 2850 Btu per horsepower may be used to estimate required flow.

Flow in Gallons Per Minute:

GPM = 2850hp / 500(temperature rise)

Assuming a 10° temperature rise, the formula may be reduced to:

GPM = .57hp

Example: Find the required radiator coolant water flow for the Ford engine if it is producing 27-horsepower at full output:

GPM = .57hp = .57(27) = 15.39 gallons per minute

Using a brief thermodynamic analysis:

Considering the engine efficiency of 26.4% and a heat rejection to the cooling water of 30%:

Btu =[(0.30) (27hp) (2545 Btu / hp)] / 0.264 = 78,085 Btu/hour

GPM = 78,085 Btu / (500 x 10° temperature rise) = 15.61 gallons per minute

15.61 GPM is close to the 15.39 GPM estimated above. The calculations appear valid.

Mixtures of antifreeze do not have the same heat transfer capacity as pure water and the flow must be increased to compensate for this decreased capacity.

Fluid Flow Correction Factors for 50% Glycol Mixtures for Heat Transfer Equivalent to Pure Water	
Fluid Temperature F	Required Flow Increase
100	1.16
140	1.15
180	1.14
220	1.14

Find the required flow for a 50% glycol mixture at 180° F:

15.61 GPM x 1.14 = 17.8 GPM

ESTIMATING REQUIRED RADIATOR AIRFLOW:

Assuming a 50° F air temperature rise, estimate the required airflow through the radiator core for the engine in the previous example. Plan for 10% system degradation over time:

$CFM = Btu\ min\ /\ .018$ *(temperature rise)*

78,085 Btu/hour / 60 min./hour = 1301 Btu/min.

1301 Btu / 50° F = 26

26 / .018* = 1446 CFM

Assuming that the system performance will deteriorate over time:

1446 x 1.1 = 1590 CFM

---

*.018 = specific heat of air .24, times the density of air at an altitude of 500 feet, .0749 = .01798

Assuming similar efficiencies and combining terms for similar formulas:

$CFM = Btu\ per\ hour\ /\ 49.1$ $CFM = 59hp$

Because Diesel engines are more efficient, a slightly lower value may be used: $CFM_{Diesel} = 55hp$

It is always wise to err on the side of excess capacity in heat transfer problems. These should be considered minimum values.

ESTIMATING RADIATOR CORE SIZE: Because we have test data from typical radiator cores, graphical solution of the radiator problem is presented here. Computational heat transfer is presented in the heat exchanger section.

Radiator cores are built with varying numbers of tube rows and fins. Cores may be described, for example, as 4 row, 9 fins per inch (FPI), 3 row 12 FPI or 1row 10 FPI. As the number of fins is increased, the heat transfer surface increases, making the unit more efficient. However it also becomes more sensitive to clogging by dirt and insects. Most automotive applications use between 10 and 14 FPI. For industrial applications, 7 to 9 FPI is more common

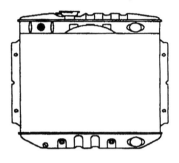
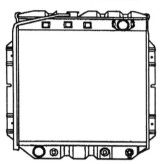

CORE: 16 3/8 X 17 1/8 X 1 1/2   ROW:3
TOP HDR: 3 1/4 X 17 3/8   INLET: 1 1/2
BOT HDR: 1 5/8 X 16 3/4   OUTLET: 1 3/4
TOC: 6 IN.   EOC: NO
STD CORE: X2307A   ROW: 3

CORE: 16 1/2 X 17 1/4 X 1 1/2   ROW:3
TOP HDR: 3.1/4 X 17 1/2   INLET: 1 1/4
BOT HDR: 1 5/8 X 16 3/4   OUTLET: 1 1/4
TOC: 6 IN.   EOC: NO
STD CORE: X2299NH   ROW: 3

Catalog Data for Ford replacement radiators. Industry #130 left and industry # 1463 right.

Looking through a radiator catalog, I find two radiators that are approximately square, downflow and fit Fords, a common make of car. Checking the index in the catalog, these fit 60's Comet, Falcon and Mustang. Hopefully, these are still common and not too expensive.

I check and see that they both have 3 rows of tubes. I now find the core area:

For radiator # 130: 16.375-inch x 17.125 inch = 280 inches^2

$$280 \text{ inches}^2 / 144 \text{inches}^2 \text{ per foot}^2 = 1.95 \text{ foot}^2$$

Now we need to see if this core area is able to dissipate 1301 Btu per minute (from previous example) with an air flow less than 20 mph or 1760 ft per minute. Checking the core data on the next page, a little less than one square foot of any 3 or 4 row core would work so either of these radiators easily has ample capacity.

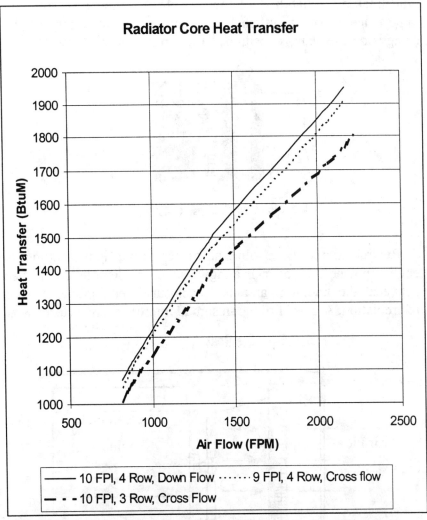

Radiator Core Heat Transfer

Water Flow = 245 lb/min 4 row exchanger, 187 lb/min- 3 row exchanger
Values are for water, not coolant mixture. Air to water temperature differential maintained at 100° F. Core area is 1 ft². Air flow velocity in FPM x 1 ft² = CFM

40. MISCELLANEOUS COOLING SYSTEM TOPICS:

Power plants may run unattended for a considerable amount of time making reliability of power plant systems of primary importance. Cooling system leaks may be may be minimized by using two hose clamps positioned 180° apart as seen below.

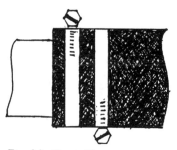

Double Hose Clamps 180° Apart

Recirculation of hot cooling air causes both a significant loss of engine power and over heating. Use of a flexible connection between the radiator and the enclosure vent prevents such recirculation. A close fitting fan shroud improves fan performance.

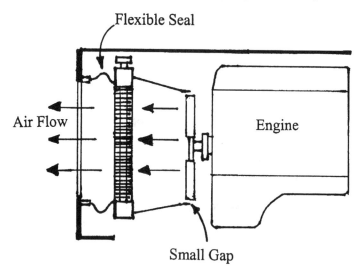

Flexible Seal

Air Flow

Engine

Small Gap

41. DESIGNING SHELL AND TUBE HEAT EXCHANGERS:

In order to minimize the explanation of the shell and tube heat exchanger design, the reader should have some background in both heat transfer and fluid flow. Since this is probably not the case for most readers, and because many are probably not interested in the calculations, I will keep section as short as possible but including enough to justify the design. Most calculations are based on the mean or average temperature of the inlet and outlet fluids, however one calculation is based on the log mean temperature difference (LMTD) of the fluids. It should be noted that I arrived at the solutions after many interations or trial calculations and that this *is* a trial and error affair as each run narrows the range of possible solutions.

Heat transfer is based on several parameters including the temperature difference between the fluids, their densities, flow rates, the viscosity of the fluids, and the turbulence in the flow. Numbers such as the Nusselt number, Prandtl number or Reynolds number describe many of these conditions. I will not define these, only give their appropriate values. Although all of the values were calculated, a few simplifying assumptions are made here: all the water jacket heat and approximately half of the exhaust heat are recoverable. The gross heat input to the generator set is 260,280 Btu. The exhaust temperature is 1215° F. Input air is 51 cfm at 70° F. Because the heat transfer coefficient of the gas side governs the size of the heat exchanger, I did not solve for the water side or the resistance of the metal tubing.

The required surface area of a heat exchanger depends upon both the amount of heat that must be transferred **q** and the heat transfer coefficient **h**.

THE EXHAUST GAS HEAT EXCHANGER:
In order to prevent condensation of sulfuric acid in the exhaust system, I will keep the final exhaust temperature at 400 °F. Although sulfur should not be a problem in propane and most

215

diesel fuels, some off spec fuels may contain enough sulfur to cause corrosion.

Example: Design a heat exchanger surface based on 3/8-inch OD tubing. The gas inlet temperature is 1215° F, the outlet temperature is 400° F. Cooling water is supplied by the engine water jacket at 160° F, the heat exchanger water outlet temperature is 185° F. The engine intake air is 51cfm at 70° F.

1. Find the LMTD

$$LMTD = \frac{(T_{gas\ out} - T_{water\ out}) - (T_{gas\ in} - T_{water\ in})}{Ln((T_{gas\ out} - T_{water\ out}) / (T_{gas\ in} - T_{water\ in}))}$$

$$LMTD = (400\text{-}180) - (1215 - 160) / \ln((400\text{-}180) / (1215 - 160))$$

$$LMTD = 532.64$$

2. Find mean temperature of gas:

$$(1215° F + 400° F) / 2 = 807.5°F$$

3. Find the volume of gas flow at the mean gas temperature:

$$Volume = (T2_{Rankin})(cfm_1) / T1_{Rankin}$$

$$(807.5 + 460)(51\ cfm) / (70 + 460) = 122cfm$$

4. Find the velocity of flow per second. Assume 24 tubes .375OD tubes with 0.035 wall thickness.

122cfm / 24 tubes = 5.08 cfm/tube
5.08 cfm / 60sec/min = .085cfs

inside area of tube = $.7854D^2$ = .7854 $((.375\text{-}.070) / 12\ in/ft)^2$

tube area = .000507 ft^2

Velocity of flow = .085cfs / .000507 = 167.5 feet per second*

*Note that this is less than Mach .2 Had this been higher than Mach .2, distortions in the flow around irregularities in the surface of the heat exchanger would change the heat transfer properties and cause a pressure drop in the unit.
Mach .2 = .2(1128fps) = 226 feet per second.

5. Find the Reynolds number (Re) at 807.5 ° F.

$$Re = \rho Vd / \mu$$

Values From Table:

ρ = density of air at 807.5° F = .0316 lb/ft^3

V = velocity = 167.5 fps

d = inside diameter of tube, ft = (.305 / 12) = .0254 ft

μ = .0805 lb/h ft, however we are working in seconds so it must be divided by 3600 second per hour.

$$\mu = .0805 / 3600 = .00002236$$

$$Re = (.0316 \times 167.5 \times .0254) / .0002236 = 6012$$

Pr = Prandtl Number = .688   k = .0299

6. Find the Nusselt number, Nu:

$$Nu = .023Re^{.8} Pr^{.33}$$

$$Nu = .023(6012)^{.8} (.688)^{.33} = 21.45$$

7. Find the heat transfer coefficient, h:

$$h = Nu\, k / d$$

$$h = (21.45)(.0299) / .0254 = 25.25 Btu /hr\ ft^2\ {}^\circ F$$

8. Find the required surface area A:

217

$$A = q / (h \ LMTD)$$

Where q = quantity of heat = approximately 15% of the gross Btu

$$q = 39,480 \ Btu$$

$$A = 39,480 / (25.25 \times 532.64) = 2.935 \ ft^2$$

9. Find the length of the tube bundle:

Surface Area of tube bundle = $\pi.0254 \times 24$ tubes = $1.915 \ ft^2$

Length of tube bundle = $2.935 \ ft^2 / 1.915 \ ft^2 / ft = 1.53 \ ft$

The length of the tube bundle is 1.53 feet*

*Note that the value varies slightly from the table due to rounding and variations in some of the starting and ending temperatures used when generating the table.

GLYCOL TO WATER HEAT EXCHANGER: The glycol heat exchanger problem is solved in a similar manor. q water jacket = 30% gross input or approximately 78,084 Btu / hr. q exhaust = 39,480 Btu / hr for a total of 117,564 Btu / hr or approximately 120,000 Btu / hr. A summary of heat exchanger solutions is included. When compared to a commercial unit, the surface areas are nearly identical; however the commercial unit claims a margin of error of ± 50%. The unit may be brought up to specification by varying the flow rates. *In order to prevent erosion corrosion of the copper tubing in the glycol heat exchanger, the maximum flow rate should be less than 5 ft per second.* Both the 18 and 24 tube bundles meet the velocity specification, however the 24 tube bundle is shorter and has less pressure drop, therefore, I will use the 24 tube bundle.

42. STRESSES IN HEAT EXCHANGERS:

Mechanical stress in heat exchangers can come from several sources including external vibration, turbulence of fluid flow, vibration at the natural frequency of the heat exchanger and

218

thermal stress. Because this is such a small unit, I will assume that it has sufficient stiffness to make the natural frequency high enough so that it is not an issue. The exchanger should be connected to the engine by a flexible connector to limit the transmitted external vibration. This leaves the thermal stress for evaluation. Stainless steel and low carbon steel have similar rates of expansion due to heating; however the copper tubing should be evaluated to see if any precautions are warranted

Coefficients of Linear Expansion k, 10^{-6} Inches per Inch per Degree Fahrenheit

Aluminum		Brass		Copper............ 9.6	
Pure............13.1		70-30..............11.1		Cupronickel..........9.5	
#356.......... 12.7		Red 85-15........10.4		Steel, mild..........5.5	
#319........... 11.1		Bronze, Silicon...9.9			

Example: Calculate the rate of expansion for 24-inch lengths of low carbon steel and copper between $32°$ F and $200°$ F.

Expansion, $E_{steel} = k\ L(\ t_2 - t_1)$ where L = length in inches

$E_{steel} = 5.5 \times 10^{-6}$ (24-inches)(200-70) = .0172-inch

$E_{copper} = 9.6 \times 10^{-6}$ (24-inches)(200-70) = .0300 − inch

At 200 degrees, the difference in length is .0128-inches

$Stress_{copper} = E\ (e/L)$ e = total change in length
E = modulus of elasticity copper, psi = 15,000,000
$Stress_{copper} = 15,000,000(.0128 / 24) = 8000$ psi

Solving for the area of copper (24 tubes) and the force on the end plate:

(8000psi)(24) .7854 $(.375^2 - .305^2 x\ in^2)$ = 7178 pounds

The cyclic loading on a restrained end plate is 7178 pounds. An expansion joint will be added.

219

Heat Transfer Properties of Air

Temperature		Cp specific heat		k		u		Density		Pr
F	C	Btu / lb F	KJ/Kg C	Btu / hr ft F	W/m C	lb / h ft	Pa s x 10^-4	lb/ft^3	kg /m^3	

Air

F	C	Btu / lb F	KJ/Kg C	Btu / hr ft F	W/m C	lb / h ft	Pa s x 10^-4	lb/ft^3	kg /m^3	Pr
0	-18	0.2401	1.005	0.0139	0.0241	0.0415	0.1716	0.0864	1.3840	0.711
200	93	0.2414	1.010	0.0184	0.0318	0.0519	0.2146	0.0602	0.9643	0.685
400	204	0.2451	1.026	0.0224	0.0388	0.0624	0.2580	0.0462	0.7400	0.683
600	316	0.2505	1.048	0.0263	0.0455	0.0721	0.2981	0.0375	0.6007	0.686
800	427	0.2567	1.074	0.0299	0.0518	0.0805	0.3328	0.0316	0.5062	0.688
1000	538	0.2631	1.101	0.0332	0.0575	0.0884	0.3654	0.0272	0.4357	0.7
1200	649	0.2692	1.127	0.0363	0.0628	0.0961	0.3973	0.0239	0.3828	0.712
1400	760	0.2755	1.153	0.0391	0.0677	0.1035	0.4279	0.0214	0.3428	0.728

Heat Transfer Properties of Liquids

Temperature		Cp specific heat		k		u		Density		Pr
F	C	Btu / lb F	kJ/Kg C	Btu / hr ft F	W/m C	lb / h ft	Pa s x 10^-4	lb/ft^3	kg /m^3	
Water										
32	0	1.0293	4.310	0.337	0.5833	4.320	17.900	62.54	1002	13.2
200	93	1.0039	4.203	0.393	0.6802	0.738	3.050	60.20	964	1.88
400	204	1.075	4.501	0.382	0.6612	0.320	1.320	53.62	859	0.91
600	316	1.525	6.385	0.293	0.5071	0.215	0.889	42.37	679	1.08
Solution of Water + 30% Ethylene glycol										
60	16	0.882	3.693	0.276	0.4777	6.040	25.00	64.90	1040	19.6
100	38	0.900	3.768	0.285	0.4933	3.270	13.50	64.30	1030	10.3
200	93	0.934	3.911	0.292	0.5054	1.230	5.08	62.10	995	3.93
300	149	0.970	4.061	0.285	0.4933	0.692	2.86	59.20	948	2.36

221

Gas to Water Heat Exchanger 39,480 Btu / hour

Tubes	Velocity in ft per sec	Re	h Btu / ft F	Area -sq ft per ft of tube bundle	Area Required	Length of tube bundle ft
12	334	11985	43.84	0.958	1.5	1.60
18	223	7994	31.73	1.436	2.09	1.46
22	182	6533	26.98	1.756	2.45	1.39
24	167	5998	25.2	1.916	2.63	1.38
26	154	5532	23.6	2.075	2.81	1.35
28	143	5133	22.24	2.234	2.98	1.33

Glycol to Water Heat Exchanger 126,408 Btu / hour

Tubes	Velocity in ft per sec	Re	h Btu / ft F	Area -sq ft per ft of tube bundle	Area Required	Length of tube bundle
12	5.475	25,271	1381	0.958	2.44	2.50
18	3.646	16,832	999	1.436	3.38	2.35
24	2.737	12,636	794	1.916	4.25	2.12

43. ESTIMATING THE LENGTH OF THE WATER HEATER LOOP:

Because 20+ percent of the typical household electrical load is used to make hot water, I added a ½-inch diameter copper tube loop inside the 500-gallon water tank to preheat the water for my water heater. This problem is solved similar to the other heat exchanger problems.

I first had to determine the water flow rate at my hot water faucet. I did this by filling a graduated bucket while timing it with a stopwatch. Converting the flow the cubic feet (1 gallon = 231 inches³, 1 foot³ = 1728 inches³).

Flow = .5ft³ / minute = 30 ft³ / hour,   62.4 lbs / foot³ x 30 ft³ = *1872 lbs / hour*

Example: Water at 72° F flows through a ½-inch copper tube with a mass flow rate of 1872 lb/ hr. The wall surface temperature is 160° F. Exit temperature of the water should be at least 130° F. Find the minimum length of the tube:

$Pr = 6.4$ m = mass flow rate   Cp (specific heat) = 1 Btu/ lbm °F
k=.35 Btu / hr-ft-°F μ (dynamic viscosity) = 2.25 lbm / ft-hr
d= diameter   q = heat transferred   h = heat transfer coefficient

Re= 4m /$\mu\pi$d = 4(1872) / 2.24 π(.5 / 12) = 25,538

$h = .023\ Re^{0.8}\ Pr^{0.4}\ (k/d)$

$h = .023\ (25,538)^{0.8}\ (6.4)^{0.4}$ (.35 / (.5 / 12)) = 1362 Btu / hr-ft-°F

q = mCp ($t_{out} - t_{in}$) = 1872 (1) (130 – 72) = 108,576 Btu

Length of tube = L = q / πd h ($t_{out} - t_{in}$)

L = 108,576 / π (.5 / 12)(1362) [160 – ((130 +72) / 2)] = 10.5 feet

The copper tubing should be at least 10.5 feet long.

44. BUILDING SHELL AND TUBE HEAT EXCHANGERS:

Both the exhaust gas and water to water heat exchangers share a similar design. They use a tube sheet on one end and a moveable tube header-expansion joint on the other. An O-ring in a machined and polished header ring seals the joint. The tube shell is made from 4-inch diameter pipe. The end caps are cast, and the header rings are also cast. The difference between the two heat exchangers is that the water to water unit uses copper exchanger tubing with cast aluminum end caps and header rings. The exhaust exchanger tubing is stainless steel; the end caps and header ring are made of cast iron. The exhaust gas exchanger also incorporates a chrome-plated piston ring to keep exhaust gas from the viton O-ring.

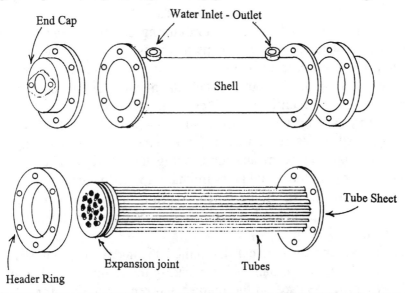

Note: **tube flow** from **hot to cold** is from tube sheet to expansion joint. **Shell flow** from **cold to hot** is from expansion joint to tube sheet.

Originally, the end cap inlets were to be tapped for 2-inch pipe thread, however this proved to be more difficult than planned and a bolt-on hose or pipe adapter was used. Because I had already made the patterns for a thermostat housing or cover, It was only a matter of pouring a few additional parts. The connection for the exhaust gas was fabricated from a pipe nipple and scrap of ¼-inch plate. Luckily, I added heavy bolt bosses to the original end cap pattern so that I did not need to make new castings.

The water inlet-outlet ports on the tube shell are made from a split 1¼-inch pipe coupling and welded into place. Turn the coupling in the lathe to make it round and remove the various casting marks, then cut it in half.

Using a torch and a circle cutting attachment, the shell flanges are cut from ¼-inch plate. They are cleaned up with a hand grinder. The bolt holes are drilled and they are welded into position. While I have welded flanges like this before, I had more trouble with distortion this time. In order to make the flanges parallel, I made a jig and turned the ends flat in the lathe. Because the jig prevents the lathe tool from cutting completely across the face of the flange, a ridge remains that must be ground off with a hand grinder as described below.

The lathe jig used to cut the shell flanges was fabricated from a section of 1 1/8-inch diameter steel rod. The ends were cut from two pulley castings with the sprue still attached. They were tapered 60° and had a setscrew in each shank. This was a very quick an dirty jig that had no means for keeping the jig tight against the work, other than the initial pressure that I applied when assembling the work on the jig. Some type of spring or screw to maintain pressure would be helpful, but involved more time than I wanted to invest making the jig.

The cast end caps are cut flat in the lathe and drilled in the mill. They are fitted to the flanges by smearing the flanges with bearing blue, rubbing the caps on the flanges and removing the high spots on the flange with a hand grinder. I made sure that the area of the flange between the shell and bolt holes had a good fit, slightly tapering the flange from the bolt holes to the outer edge. This ensures that the inner edge of the flange makes contact with the gasket first.

226

Weldable pipe flanges are available and probably worth investigating. 3/8-inch plate would also be less likely to warp and have more material available when cutting the flanges parallel. The finished shell is coated inside and out with Ospho-rust protectant. After a few days, any excess is cleaned up with a Scotch-Brite pad and the shell is given a coat of ACE hardware rust stopper enamel.

Gaskets for the water to water heat exchanger are cut from a 50-durometer-hardness sheet of 3/32-inch thick Buna N rubber from MSC Tools. I had originally intended to cut the exhaust gaskets from "Kao wool felt," however I was given a scrap sheet of graphite gasket material while having a propane hose made at a hydraulic shop. Some boiler supply houses may call the Kao wool "boiler paper or felt."

The tube sheets were made from ¼-inch stainless plate and 3/32-inch brass sheet, depending upon the application. The thickness of the material is not so important as is the availability. Prices for the material varies considerably. One place wanted $50 per piece for a 1-inch length of 4¼-inch diameter stainless rod, while another shop said $5 for as much as you can haul off out of the scrap bin. The brass plate was also found in their scrap bin.

3/8-inch OD #304 stainless tubing, .035 wall, was purchased from McMaster Carr. 3/8-inch OD copper was purchased locally. Note that ¼-inch nominal diameter copper tubing has a 3/8-inch OD. Copper is also available in various wall thickness'. Type K = .035 wall, L = .030 wall and M = .025 wall for ¼-inch diameter tubing. Type L was available locally, however hoping for a longer tube life I would have preferred type K.

In an effort to keep the tube lengths somewhat standard, I have deviated from the actual calculated dimensions. The tubes are cut into 2-foot lengths. Shortening the tube shell so that the tubes protrude about 1/8 to ¼-inch on each end of the assembly (accounting for gaskets). The tubes are slightly flared on one end so that they will press into the tube sheet. In commercial headers, tubes are often flared while in the tube sheet. The cast iron header ring is assembled (without O-rings or piston rings) and the tubes are TIG welded into position. Again using the cast iron ring, the copper tubes were brazed, however because of the lower water temperature, they could have been soldered. A jig to keep the plates pushed out during assembly might help. Next time I might add a section of 3/8-inch tubing over a tube in 4 opposite corners to keep the plates pressed out evenly. This would become a permanent part of the heat exchanger. Threaded rod and a few nuts might be a better answer.

A 4-inch diameter pipe is actually 4.026 inches ID, however the weld where the pipe was made protrudes inside the pipe an additional .015 inches. A small relief must be filed into the expansion-joint tube head in order to pull it back through the shell.

Because I did not have a 4-inch diameter copper or bronze rod, the expansion joint head for the copper tubing is cast from silicon bronze. The diameter is cleaned up and the diameter is made as closely round as possible on a disk sander then the part is moved to the lathe where the faces are cut flat. The part is centered on the mill and the tube holes are drilled. The part is then returned to a jig in the lathe and the OD and O-ring grove are both cut.

The jig is fabricated from a section of 2-inch pipe welded to a plate left over from cutting the flanges. The part is turned concentric in the lathe. Two holes are located, drilled and 3/8-inch diameter pins are inserted and tack welded into position. The pins are smoothed with a file.

The diameters of the expansion joint tube heads are cut .003 small for an OD of 3.997. The maximum specified clearance between the male and female parts for proper operation of the O-ring is .006 or a diameter of 3.994. Parker claims the seal should be good for up to 1500 psi. The maximum system pressure is on the engine side, where the header tank has a 12-13 psi. radiator cap, so we should be good here.

Viton has a working temperature of + 400°F and up to 600°F for brief periods. The expansion-joint O-ring is a Viton, 4-inch OD, 3.374-inch ID, .139 wide Parker # 240. These were inexpensive and I purchased a bag because I knew I would probably shred one or two fitting the parts. The critical feature of the assembly is the groove depth or groove OD. The groove OD is 3.778-inches. This forces the O-ring out against the sides. If the groove is too deep, the O-ring does not fit tightly. If too shallow, it can not be assembled. Smear the ring and header ring with petroleum jelly and test the fit as you are cutting the groove.

The piston ring is compressible and the fit is not as critical. I cut the groove diameter .01 small, the final groove OD being 3.630 – 3.635-inches. The piston ring is available from Hastings, part # 26599. The 1/8th inch ring fits a 4-inch diameter cylinder. The ring is flash-chromed for corrosion protection.

Over-Temperature Protection: A temperature switch should be installed in both the exhaust heat exchanger and engine and connected to the engine protection relay of the engine controller.

NAPA #	Temperature	Thread
TS 6668	230F	3/8 x 18
TS 6642	240F	3/8 x 18
TS 6665	257F	3/8 x 18

There are several possible flow paths for the cooling water. Initially, the exhaust heat exchanger was to utilize engine jacket water so that there would be no condensation in the heat exchanger. However, since it is made of stainless steel, it doesn't matter if the exhaust is wet.

The 160°F thermostat was selected primarily as a good temperature for water to be *entering* the engine water jacket. If the thermostat is used on the engine *outlet*, then the temperature is not critical and may be +200°F.

Two flow paths are described here. The first is featured in the generator project overview chapter and is described as system #1. System #2 requires a little more piping but provides faster engine warm up and maintains coolant flow through the exhaust heat exchanger. I will eventually switch to flow path #2.

System #1. See drawing on page 63. Because it is the simplest, I am using this system without an additional thermostat. A thermostat may or may not be used in the exhaust heat exchanger.

In order to ensure a flow of water, even when starting and the engine is cold, a bypass hole is drilled in both the thermostat at the exhaust heat exchanger and at the engine.

If used, the shell thermostat is attached with a fabricated housing. A plate is welded to a pipe nipple and a thermostat cover is cast. This is not particularly critical and any thermostat cover that fits may be used...something else to pickup at the scrap-yard when looking for radiator necks.

Asking for a thermostat by temperature baffles many counter salesmen however, a 160°F thermostat is available from NAPA as thermostat part #THM 70. Several thermostats use this part number therefore, you must also specify the temperature. The flange diameter is 2.125-inches. A heavy-duty 2.125-inch thermostat is #THM 530060. Part #THM 91 has a 2.5-inch flange. THM 532060 is a heavy-duty 2.5-inch flange. I am sure there are others available

Currently, the coolant is moved through the water to water exchanger with a ½-hp electric pump. Although I have not done it yet, I plan to add an automotive fan switch to the water to water heat exchanger. This, combined with a relay, would turn the pump on and off as required to maintain a constant temperature. NAPA fan switch part # for 122°F = FS 169. This is a metric size and fits Toyotas. Another part number for higher temperatures is Part # FS 142 for 165°, this is a two step switch. One turns on at 165, the other at 179° F.

System #2. Because one channel is always open, providing coolant flow through the exhaust heat exchanger, this is probably the better system, however it requires fabrication of a small bypass manifold. A 3/8-inch heater hose connects the bypass to the exhaust heat exchanger.

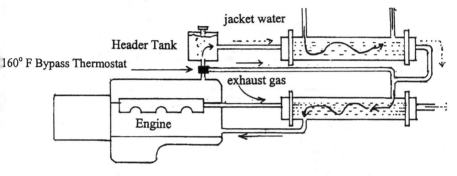

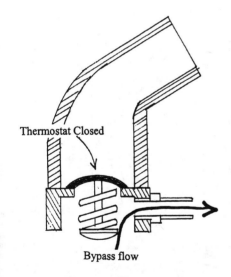

Above: Bypass Thermostat

Assembly of the heat exchangers is straightforward. However, compression of the piston ring requires a small clamp. A 4-inch hose clamp may be used but requires twisting a screwdriver for a while. I used a ½-inch wide strip of sheet metal with two tabs bent up at 90°. These are pinched with a vise grip or pliers while the piston ring slides into the header ring. Because of the high operating temperature of Viton, it is quite possible that the piston ring could be deleted and two O-rings used.

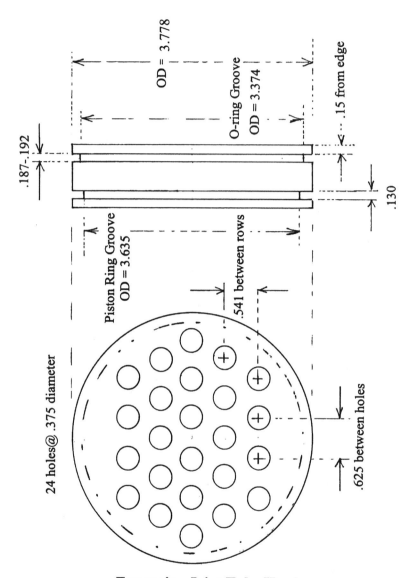

Expansion Joint Tube Header

Note that **tube sheet** on opposite end of the tubes is 7 inches in diameter and uses the same tube hole pattern as seen above. The bolt holes are 3/8th inch diameter and the bolt hole pattern is the same as the header ring. Material is ¼ inch thick or less. See text for details.

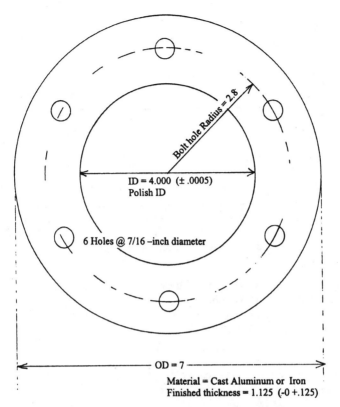

ID = 4.000 (± .0005)
Polish ID

6 Holes @ 7/16 –inch diameter

Bolt hole Radius = 2.8

OD = 7

Material = Cast Aluminum or Iron
Finished thickness = 1.125 (-0 +.125)

Above: **Header ring** detail. Note: The **Shell Flanges** are identical except they are made from ¼ to 3/8-inch steel plate, the ID = 4.5 inches and the bolt holes are 3/8 inch diameter. The **Tube Sheet** mates with the flange and shares the internal hole pattern with the expansion joint on the previous page. The end caps are secured with 5/16-ich bolts. Below: **Flange Turning Jig** One end is clamped in the 3 jaw chuck the other is held by the lathe tailstock.

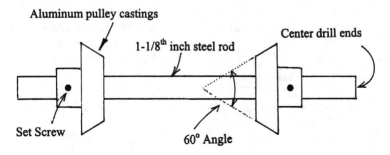

Aluminum pulley castings

1-1/8th inch steel rod

Center drill ends

Set Screw

60° Angle

233

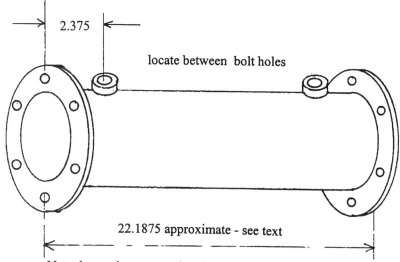

2.375

locate between bolt holes

22.1875 approximate - see text

Note: locate the ports so that they do not block the bolt holes.

Right: Expansion joint turning tool. See Text. The small piece is a 1-inch diameter steel rod that has been cut flat , parallel then center drilled. A live center in the tailstock is used to press the tube head against the jig.

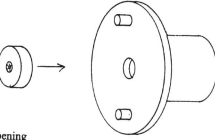

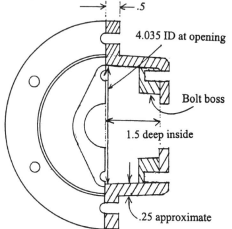

← .5

4.035 ID at opening

Bolt boss

1.5 deep inside

.25 approximate

Left: The **end cap** pattern is made from built up sections of plywood and then turned in the lathe. ¾-inch diameter bolt bosses are added under the 2-bolt flange. Castings are made from aluminum or cast iron depending upon application.

234

45. NOISE AND VIBRATION:

Noise and vibration are related and occur in waves similar to electric waves. Electric waves can be converted into vibration and noise by sending them through a loudspeaker. Similarly noise and vibration may be converted into electric waves by the use of a microphone. Many of the terms used to describe electric waves may be used to describe sound waves.

Frequency is the number of complete cycles of an alternating waveform measured in time such as seconds or minutes. The unit of measurement is Hertz (Hz), if the time is in seconds. 4 Hz being

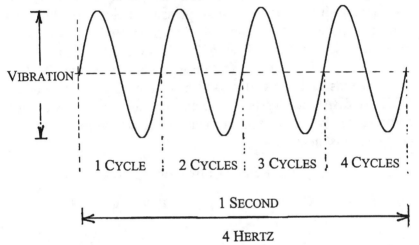

4 cycles per second. 60 Hz being 60 cycles per second. Vibration over 20 Hz produces audible sound with lower frequencies producing lower tones and higher frequencies producing higher tones. The highest audible tone is between 15,000 and 20,000 Hz, depending upon how good your hearing is.

Lower frequencies travel around corners, through walls and holes. High frequency noise is easily reflected or absorbed. Because they have different properties, different methods are used to control high and low frequency sounds.

235

For engineering purposes, the range of frequencies is divided up into octave bands. The highest tone in an octave band is twice the frequency of the lowest tone.

Octave Band Center Frequencies, Hz

63   125   250   500   1000   2000   4000   8000

Besides frequency or pitch, we also hear amplitude or loudness. Loudness is measured in decibels or dB. Because we can hear over a very wide range, the dB scale is logarithmic. The lowest sound we might hear is 1 to 10 dB and another sound might be 100 dB. The actual difference in sound pressure being 100,000 times greater than the faintest sounds that we hear. Because our ears are most sensitive in the 500 to 6000 Hz range, the dBA scale is corrected for our hearing. The dBA scale is the sum of all the octave bands and is usually used in community noise standards. While a dBA reading on a meter may tell you if you are in compliance with a local ordnance, it is useless from an engineering standpoint.

Octave Band Correction for dBA

63	125	250	500	1000	2000	4000	8000
-26	-16	-9	-3	0	+1	+1	-1

Because the dB scale is logarithmic you can not directly add decibels. Addition of decibels is done by adding a correction factor to the higher decibel sound.

Addition of Decibels

Difference Between Sound Sources	Correction Factor to be Added to the Higher Decibel Source
0 or 1 dB	3 dB
2 or 3 dB	2 dB
3 or 4 dB	1 dB
9 db +	0 dB

Example: Add the octave bands produced by a diesel generator set to find the dBA:

	63	125	250	500	1000	2000	4000	8000
	Octave Band dB at 10 feet from Generator							
Band dB	91	98	91	85	86	82	80	65
dBA Correction:	-26	-16	-9	-3	0	+1	+1	-1
Corrected dB:	65	82	82	82	86	83	81	64

Addition of Decibels using the Addition Correction:

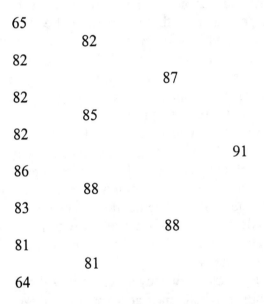

```
65
        82
82
                87
82
        85
82
                        91
86
        88
83
                88
81
        81
64
```

The diesel set noise is 91dBA.

Sound pressure levels should also specify the distance from the source. Octave sound pressures are typically measured by a frequency analyzer with a graphic display. These may be found locally at music stores that sell sound reinforcement equipment or PA gear. (I rented one for $10 for the afternoon, thanks George!)

VIBRATION:

Both generators and engines are sources of vibration. Any imbalance in a rotating body causes a displacement at a frequency relative to both the speed and amount of imbalance. Vibration occurs at the rotational speed and at harmonics or multiples of that speed. For example an 1800 rpm set generates a fundamental frequency of 1800 / 60 seconds = 30 Hz. Harmonics also occur at 60, 90, 120 Hz and so on.

Vibration isolators are used to reduce the amount of vibration that is transmitted to the foundation of the generator set or building. Vibration from a generator set that is directly bolted to a concrete slab will soon crack and beak it. Spring or rubber type isolators must be used between the generator and the foundation. Likewise, flexible connections must be made between the engine and any other system such as the exhaust system, cooling system or air input. Hard or rigid mounts are never used.

Different types of vibration isolators are used for different frequencies. Rubber type isolators are good for higher frequency isolation and spring type mounts are best for low frequency vibration isolation. Often, both types are combined such as a rubber pad under a spring type isolator. The degree of isolation is related to the amount of static deflection of the isolator with larger deflections giving greater amounts of isolation. Static deflection is the amount of deflection when the system is at rest.

Each system will have a natural frequency at which it will ring or *amplify* the vibration input. A bell rings at a specific tone or frequency. This particular tone is the bell's natural frequency. Harmonics are generated in the bell relative to the bell's shape, composition and damping. At a frequency above the natural frequency, there is a crossover point where that ratio of vibration that is transmitted to the system equals 1 or all of the vibration is transmitted (however, not amplified). This crossover point is 1.414 times the natural frequency of the system regardless of damping.

As the frequency increases further, the vibration is attenuated with the degree of attenuation increasing with the increasing frequency.

Damping or friction may be added to the system to reduce the amplification at the natural frequency, however damping increases the amount of high frequency vibration that is transmitted. For any system without damping, the static deflection is related to the natural frequency by:

$$\text{Deflection} = 9.8 \, / \, \text{Natural Frequency}^2$$

$$\text{Natural Frequency} = \sqrt{9.8 \, / \, \text{Deflection}}$$

Example: What is the natural frequency of a spring system with a deflection of .375 inches?

$$\text{Natural frequency} = \sqrt{9.8 \, / \, .375} = 5.11 \, Hz$$

Natural Frequency is also determined by:

$$\text{Natural Frequency, Hertz} = 3.13 \, \sqrt{K/W}$$

$$\text{Natural Frequency, cycles per minute} = 188 \, \sqrt{K/W}$$

Where K = System stiffness   W = Weight of isolated system

These relationships are plotted on the next page for selection of natural frequency, deflection and vibration isolation for undamped (spring type) isolators. 30% Transmission of vibration is considered the maximum for non-critical systems. 10% is the maximum for critical applications and 1% is for extremely critical situations.

Example: Select an isolator for an 1800 rpm diesel generator set to be used in a residential location. The weight of the diesel is 792 pounds and the generator weighs 462 pounds. The diesel and the generator each have 4 mounting feet.

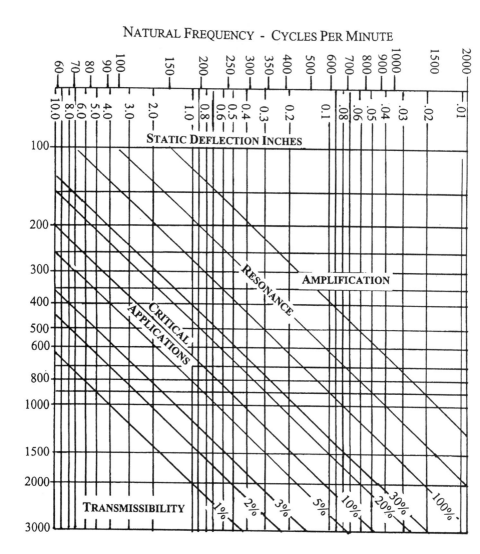

The natural frequency of the diesel set is 1800 / 60 = 30 Hz. This is deemed a critical application because it is located in a residential area. 10% maximum transmissibility is selected. A horizontal line is extended from the rotational frequency of 1800 to the 10% line.

Moving vertically, the natural frequency should be less than 500 cycle per minute, or 500 / 60 = 8.33 Hz.

Although the engine and generator bolted together, I will estimate the springs separately because one end is much heavier than the other. Using the springs we found in the spring chapter, the spring rate is approximately 217 pounds per inch.

Assuming one spring for each foot of the diesel, 4 springs x 217 lbs/inch = 868 lbs.

Diesel weight of 792 lbs,

792 lbs / 868 lbs/inch = .912 inch deflection

This is too much deflection. The spring coils will close. I will try doubling the number of springs for total of 8.

792 lbs / 1736 lbs/inch = .456 inches deflection

This is an acceptable amount of deflection. Now for the generator end using 4 springs:

462 lbs / 868 lbs/inch = .532 inches deflection

There is less than a tenth of an inch difference between these and when the unit is bolted together, it may level out a little more. Now to check the natural frequency of the average deflection, .494 inch.

Natural Frequency = $\sqrt{9.8 / .494}$ = 4.45 Hz

Checking the crossover point:

4.45 x 1.414 = 6.3 Hz

Both of these frequencies are below the critical 8.33 Hz Checking the chart on the previous page, .5 inch deflection at 1800 rpm gives us a *very good* 2 ½% transmissibility, making for an excellent vibration isolator.

46. MAKING VIBRATION ISOLATORS:

Spring type vibration isolators are not difficult to make. These isolators are made from discarded valve springs, 3/8 by 2-inch hot rolled flat steel, and steel rods ½ and ¾ inch diameter.

A trip to the local engine rebuilder yielded a bag of discarded valve springs. These were sorted by size and checked for spring rate as described in the spring chapter. After determining the springs were suitable in the design section, the isolators were fabricated as follows.

Note that the lengths of the rod used here are suitable for the calculated deflection. Depending upon your springs and the weight of your engine generator, you may have to shorten the rods to prevent them from hitting the bottom of the mount. Lighter sets may require only 1 spring.

Make the spring center post from ¾ inch steel rod. To be sure that the center hole is concentric, drill the rod in the lathe and chamfer both the inside and outside diameters. The ½ inch in diameter rods slide freely through the center holes to keep the set aligned. However there will be some distortion during welding, so the center holes must be drilled oversize.

The holes in the flat bar must be drilled accurately or the assembly will bind and be useless as an isolator. Provide a good chamfer on the side of the flat bar to be welded. The weld bead will fill this chamfer and be ground down to make the surface flat.

Press the rods into their respective holes and without using the springs, set the top and bottom sections together for welding. Tack the rods into position using many small welds to minimize the distortion. Label the tops and bottoms so that they may be reassembled as matched sets. Remove any burrs and check for smooth operation. Some of my isolators were a little stiff however they loosened up after a few hours use.

The isolators are painted and greased before assembly. Because of the difference in height of the motor mounts and the generator mounts, a steel pipe riser is used on the front mounts as seen in the

242

generator photo. The engine and generator mounting bolts run in oversized holes and have small springs or rubber washers on top. The bolts are not tightened but are drilled for cotter pins at the bottom and a nut is run up just over the cotter pin hole.

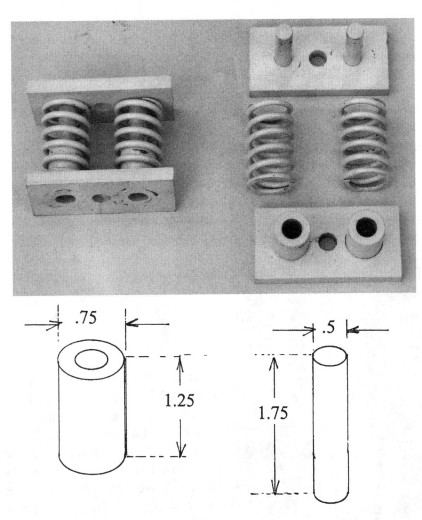

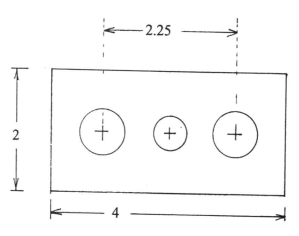

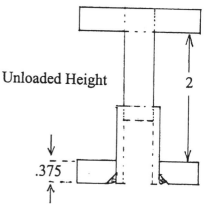

Unloaded Height

2

.375

Left: The 2 inch diameter pipe is notched 1.75 inches wide and 1.5 inches deep for access to the bolt. This particular bolt is secured with a lock nut and some lock-tite thread sealer. These nuts readily vibrate loose and fall off. Later versions skipped the upper spring and the lower lock nut. The bolt was drilled for a cotter pin.

47. PROPERTIES OF SOUND:

Solution of acoustic problems is fairly math oriented; however by understanding a few properties of sound, practical solutions yielding considerable improvement in the situation are readily available.

Because we hear over such a wide range of sound pressures, the problem in noise control is that much noise reduction is required to make a noticeable difference. A reduction of 10dB is perceived as about half as loud as the original sound level, however it requires a 90% reduction in the sound pressure level. Therefore to achieve a large amounts of noise reduction, very large amounts of sound energy must be removed. To reduce the loudness of a sound to ¼ of its original value, 99% of the sound energy must be removed. This corresponds to a 20dB reduction in noise level. A 30dB reduction requires 99.9% reduction in the noise level.

Sound Power of Various Sources		
Source	Power	dB
Jet aircraft	10,000 Watts	160
Pneumatic Hammer	1 Watt	120
Automobile 45 MPH	.01 watt	110
Piano	.002 watt	103
Conversational Speech	.000,002 watt	73
Soft Whisper	.000,000,001 watt	30

SOUND OUTDOORS: As sound radiates from the source, the pressure is spread over a larger and larger area like ripples in a pond after a stone is dropped into the water. Mathematically, doubling the distance decreases the sound pressure by 6 dB, however because of ground reflections; a more realistic figure is 5 to 5.5 dB. The distance from the source is always included in any sound pressure measurement.

MASS LAW AND FREQUENCY: The weight of a partition has an effect on it noise reducing ability. The amplitude or loudness of the sound radiated from a wall is relative to the amount of vibration in the wall, which is in turn governed by:

$$Force = Mass \times Acceleration$$

Increasing the mass of the wall reduces the acceleration or vibration of the wall. Doubling the weight of the wall reduces the sound transmission by about 5 to 6 dB. Doubling the frequency has the same effect as doubling the weight of the wall. You can expect about a 5 to 6 dB reduction in sound.

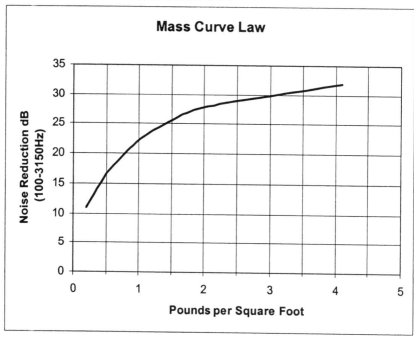

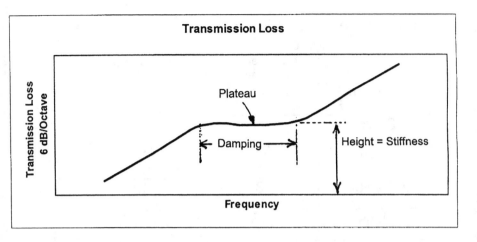

Transmission Loss

The transmission loss or noise reduction for various materials may be estimated by:

$$\text{Noise reduction} = 20 \log (W) + K$$

W = Weight of material in pounds per foot^2
K = Frequency dependent noise reduction factor from the table

K values dB for Octave Bands:

Octave Band Center Frequencies, Hz

	63	125	250	500	1000	2000	4000	8000
K, dB	3	9	15	21	27	33	39	45

Note that when you reach the plateau height, the noise reduction will remain the same for the next three octave bands. Above the plateau, the noise reduction will be 6 dB per octave.

Density and plateau heights are given in the following table:

247

Surface Density and Plateau Heights		
Material	lbs./ square foot per inch	Plateau Height dB
Aluminum	14	29
Brick	11	37
Concrete	12	38
Fir Plywood	3	19
Glass	13	27
Lead	59	56
Plaster	9	30
Steel	40	40
Cinder Block	32	33

Note that cinder blocks are weight per square foot of surface area.

Example: Estimate the transmission loss for 1/16-inch thick steel.

First, find the weight of 1/16-inch thick steel per foot^2

Weight = (40 lbs./inch thickness) / 16 = 2.5 lbs.

1/16th inch steel = 20 x log (2.5) = 7.96 dB or approximately 8 dB

The K factor for the octave bands is found on the previous page:

Solving for the first Octave band 63Hz:

Noise reduction = 8dB + K of 63 Hz

Noise Reduction = 8 dB + 3 dB = 11dB

Solving for the rest of the octave bands:

Octave Band Center Frequencies, Hz

	63	125	250	500	1000	2000	4000	8000
dB	11	17	23	29	35	40	40	40

Notice that when we reach 40 dB we are at the plateau and the last 3 octaves remained the same.

Two-inch thick plywood has less noise reduction than 1/16-inch steel.

HOLES OR LEAKS IN ENCLOSURES: The effectiveness of any enclosure is considerably reduced when it has openings. One square inch in a 100 square foot wall will transmit as much sound as the entire wall. Often, soundproof enclosures are sealed. This does not mean that there is no airflow. Cracks and openings are gasketed, like an auto door. Airflow openings are lined with sound absorbent material and have one or more 90° bends.

DIRECTIVITY DUE TO PIPE DISCHARGE OR OPENING: When sound is emitted through an opening such as an exhaust pipe, it may not radiate uniformly in all directions. If the wavelength of the sound is shorter than the opening, most of the sound goes in the direction of the opening. Because higher frequencies have shorter wavelengths, higher frequencies are more directional.

To find the wavelength of a frequency:

Wavelength = 1128 / frequency

Find the wavelength at 4000hz:

1128 / 4000 = .282 foot   .282 foot x 12 inches / foot = 3.384 inches

Frequency = 13536 / L where L = diameter of opening in inches

PANEL DAMPING: The transmission loss of metal or wood panels can be significantly increased by the application of damping material. Rubberized or viscous damping material may be applied by spray, trowel, or in the form of sheets. Damping increases proportionate to the square of the thickness ratio of the damping material to panel thickness. Usually 1/1 or 2/1 is sufficient. Damping materials are temperature dependent. There is a temperature of maximum damping, above and below which the damping decreases rapidly.

ENGINE NOISE: While adding a proper muffler quickly reduces the exhaust noise you will find that the overall engine noise of the generator set is not much improved. Several sources contribute to engine noise. They include mechanical noise, combustion pressure, fan noise, air intake and exhaust noise.

Exhaust noise varies considerably between engines depending upon valve timing and increases about 15 dB from no-load to full load. Exhaust noise is typically in the 50 to 200 Hz range and easily reduced with a reactive type muffler. For critical applications, a second absorptive silencer is added to attenuate mid to high frequencies (500 to 4000 Hz).

Because diesel engines are unthrottled, intake noise is much louder than that of gasoline engines. The fundamental frequency of intake and exhaust pulsation is:

Frequency = (rpm x no. of cylinders) / 120

Adding a muffler to the intake reduces the inlet pulsation of a diesel engine. The muffler must be of the reactive type and unlined or having no fibers or sound absorbing material that can be sucked into the engine. However, no input muffler may be required with a good enclosure. Turbo noise is typically in the 2000 to 4000Hz range and may have a muffler or is attenuated in the generator enclosure.

By doubling the rpm, mechanical noise increases by approximately 12 dB. 1800-rpm sets are much quieter than 3600-rpm sets.

FAN NOISE: Engine fans generate considerable noise with the frequency being dependent upon both the type of fan, rpm and the number of blades. Although there are may types of fans, propeller and vaneaxial fans **(right)** are most common and are considered here. To estimate a fan's dB level, a calculated dB level is added to a basic sound power level for each octave band. The fundamental frequency of the fan blades is calculated and the dB correction is added to the corresponding octave band.

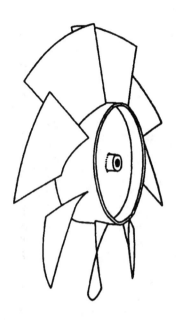

Fan Noise Level* per Octave = B_L + 10 Log F + 20 Log P

*at 1 inch WC pressure

B_L = Basic octave level from chart,   F = CFM   P = pressure, inches WC

Blade Fundamental Frequency = (rpm x no. of blades) / 60

Example: Find the dBA rating of a four blade propeller type fan running at 2200 rpm producing 3000 cfm at 1.125 inch water column

10 Log (3000) + 20 Log (1.125) = 35.8 dB or approximately 36 dB

Blade Fundamental Frequency = (2200 x 4) / 60 = 147 Hz

Using the chart of Basic Noise Levels for 125 Hz:

Noise Level dB

Fan Type	Blade Tone Correction dB	Octave Band Center Frequency							
		63	125	250	500	1000	2000	4000	8000
Vaneaxial	8	42	39	41	42	40	37	35	25
Propeller	6	51	48	49	47	45	45	43	31

Noise @ 125 Hz = 48 dB + 36 dB + 6 dB (blade tone correction) = 90 dB

Completing the rest of the octave bands:

Octave Band Center Frequencies, Hz							
63	125	250	500	1000	2000	4000	8000
dB 87	90	85	83	81	81	79	67

Correcting for dBA:

Octave Band dB for Propeller Fan							
63	125	250	500	1000	2000	4000	8000
Band dB 87	90	85	83	81	81	79	67
dBA Correction: -26	-16	-9	-3	0	+1	+1	-1
Corrected dB: 61	74	76	80	81	82	80	66

Adding per rules for addition of decibels:

```
61    74    76    80    81    82    80    66

   74          81          85          80

      81                      85

            86
```

The fan noise is 86 dBA at the enclosure

PROPERTIES OF ACOUSTIC MATERIALS: Acoustic materials include sheets and viscous coatings for damping, fiberglass and foams for absorption, and composites of foams and damping material. While there are many other materials, these are of primary importance for engine compartments.

The cellular structure of the foam dictates its acoustic properties. Not all foams are the same. They vary in both the size of the pores and the percentage of pores that are open or closed. As the sound waves move through the porous material they are converted into heat by the resistance to flow. If the porosity is high with large open cells, then the sound waves move through with little absorption. If the cells are closed, then the sound will bounce off and again there will be little noise reduction. A good foam structure for sound absorption has approximately 70% of its cells open and 30% closed. Changing both the cell size and percentage of closed cells changes the frequency response of the foam. A lower percentage of closed cells has a higher frequency absorption, which is desirable because higher frequencies are both more annoying and damaging to the hearing. Smaller cells are better than larger ones. Thickness of the material is also important, with foams ½-inch or less being of little value. Urethane foam is typically used for sound absorption and weighs about 2 lbs. per ft^3.

Closed cell foams are poor sound absorbers, however they are good heat insulators and are typically used as decouplers or vibration isolators or yoga mats. These foams are made of polyethylene or PVC. With the PVC foams being both much more heat resistant and less sensitive to organic solvents – gas and oils. Your wife's yoga mat is made from polyethylene.

Because foams are absorbent materials, they are likely to become contaminated with dirt, oil, fuel or water, closing off the cells and destroying their sound absorbing properties. Which, by the way, is why you never paint sound absorbing materials. Several methods are used to prevent contamination and are often based upon a thin limp plastic film. The film must be thin enough to allow the sound to pass through and the adhesive used to attach the film must not close off too may cells. Adhesive should be applied in spots or strips. The foam may also be enclosed in a bag, which eliminates the adhesive problem. Adding a film or bag changes the frequency absorption nature of the foam to that of a more closed cell foam. PVC (vinyl) films are better suited to engine compartments that polyethylene. Aluminized Mylar or polyester films are also available. Films usually increase the low frequency absorption at the expense of high frequency absorption.

Foams may be attached with adhesives or mechanical fasteners. Adhesives should be resistant to both heat and oils. Adhesives that are thinned by mineral spirits are not suitable for engine compartments. Spacing the foam away from the reflective surface by ½-inch or more considerably increases the sound absorption properties of the foam.

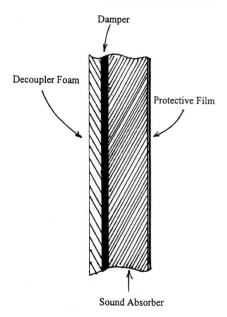

Usually noise problems require more than just one type of material and absorbers when combined with dampers are more effective. These may be applied separately or as a composite material (**Right**).

SOUND ABSORPTION COEFFICIENT: The sound absorption coefficient α is the percentage of sound energy directed at 90° to the surface that is absorbed or dissipated. When selecting a foam or fiberglass absorber, it should have a sound absorption coefficient of at least 50% in the frequency range of interest.

TRANSMISSION LOSS: Transmission loss refers to the loss in dB as the sound moves through a barrier such as a brick wall. Composite foams may also be rated by transmission loss.

Transmission Loss dB for Solid 8-inch Concrete Wall

Hz	125	500	2000	4000
Loss dB	36	43	50	61

Absorption Coefficient - Typical Acoustic Foam

Foam Thickness Inches	Frequency Hz						
	125	250	500	1000	2000	4000	NRC*
1/2	0	.08	.16	.55	.98	.95	.45
3/4	.01	.11	.38	.92	.93	.86	.6
1	.06	.19	.74	1	.85	.97	.7
1-1/2	.1	.3	.77	1.04	.99	1.11	.8

* ASTM C-423-90A Absorption Coefficient

Absorption Coefficient - Aluminized Film Acoustic Foam

Foam Thickness Inches	Frequency Hz						
	125	250	500	1000	2000	4000	NRC*
1/2	.08	.15	.53	.23	.15	.28	.35
3/4	.08	.33	.87	.38	.29	.32	.45
1	.11	.56	.61	.31	.34	.46	.45
1-1/2	.2	.75	.49	.37	.31	.49	.5

* ASTM C-423-90A Absorption Coefficient

Transmission Loss in dB - 1-3/8 inch Composite Foam

	Frequency Hz						
	125	250	500	1000	2000	4000	*STC
Foam	15	17	19	26	37	50	24
on 16 ga Steel	23	26	26	31	50	62	32

*Sound Transmission Class

LINED DUCTS AND BENDS: "Sealed" generator enclosures must be vented. Propeller type fans, typically used in radiator applications, produce high flow rates and low static pressure. Because these fans produce such low pressure they require noise reduction with very little resistance to airflow. Ducts and bends are lined with sound absorbing material for these situations. The attenuation of lined ducts is estimated by:

Attenuation = $4.2\alpha^{1.4}$ x (Length / Diameter)

For rectangular ducts: Diameter = 4(sectional area / perimeter)

Because the attenuation is low, very thick linings are used, often in combination with splitters.

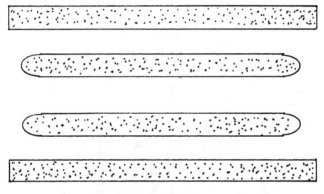

Duct with Splitters

The addition of splitters can greatly improve the performance of the attenuator however they create a greater pressure drop. Data for commercial attenuators usually includes flow and pressure drop as well as octave band attenuation.

Absorptive linings must be securely attached either by mechanical fasteners or glue. The glue must not degrade the performance of the lining. Hardware cloth and perforated sheetmetal are used for increasing flow rates. The flow should be below 1750 fpm to avoid generating noise at the outlet.

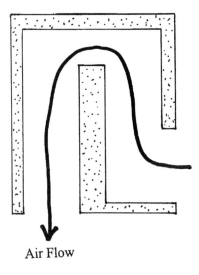

Air Flow

Higher loss devices have no line of sight between the inlet and outlet. The more times the sound is reflected, the higher the attenuation. All internal surfaces are treated with absorbent material. Mid and high frequencies may be attenuated with lined ducts. Peak frequencies of attenuation are usually 500 to 2000 Hz, depending upon the liner material. Attenuation of frequencies below 500 Hz drops off rapidly. Liner thickness may be increased for greater attenuation at the expense of higher static pressure loss.

Air Flow

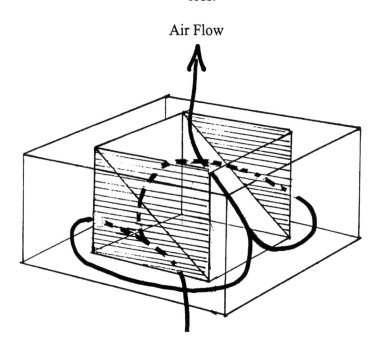

Attenuation of 36" Lined Square Ducts, dB*

Duct Width Inches	Frequency Hz							
	63	125	250	500	1000	2000	4000	8000
12	0	1	4	11	21	18	9	6
14	1	3	6	12	19	11	9	6
16	2	4	6	12	17	13	9	7
18	2	4	7	12	15	12	9	7
20	4	7	13	22	29	16	12	10
24	2	4	7	11	13	9	8	7

*Kinetics - Vibron #VSR-4   Flow = 1500 fpm

Attenuation of 60" Lined Square Ducts, dB*

Duct Width Inches	Frequency Hz							
	63	125	250	500	1000	2000	4000	8000
12	1	2	8	21	44	31	12	10
14	3	5	12	24	39	26	13	10
16	3	5	12	25	35	22	13	10
18	3	6	12	27	30	20	12	10
20	4	7	13	22	29	16	12	10
24	6	9	14	22	22	15	12	12

*Kinetics - Vibron #VSR-4

Pressure Drop of Lined Square Ducts*		
	Flow fpm	
Pressure Drop Inches WC	36 inch length	60 inch length
.08	1360	1180
.10	1520	1320
.12	1660	1440
.14	1790	1560
.16	1920	1670
.18	2030	1770
*Kinetics - Vibron #VSR-4		

48. MUFFLERS AND EXHAUST SYSTEMS:

Mufflers may be divided into three types, reactive, dissipative and a combination of the two. Reactive mufflers are tuned to the desired frequency where they are very effective. At other frequencies they have no benefit and at some frequencies they make things worse. Dissipative mufflers are untuned frictional devices that convert the acoustical energy to heat.

Reactive mufflers use a sudden increase of internal area with increasing attenuation occurring with increasing expansion.

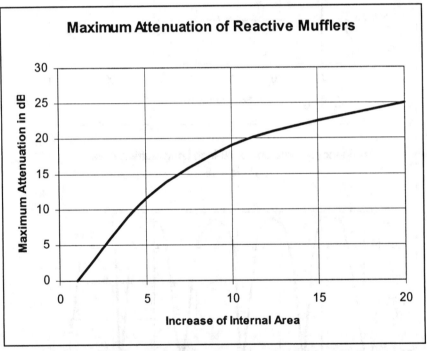

Maximum attenuation occurs when the length of the muffler section is equal to ¼ wavelength. The frequency of maximum attenuation is:

Frequency (f) $= 1128 / 4 L$ or $L = 1128 / 4f$

L = length of muffler section in feet

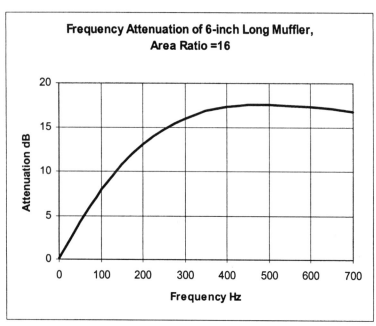

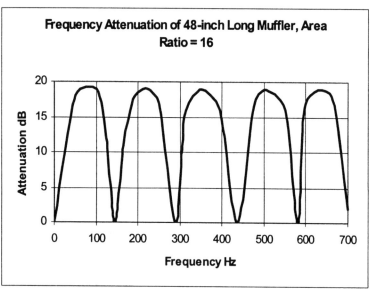

262

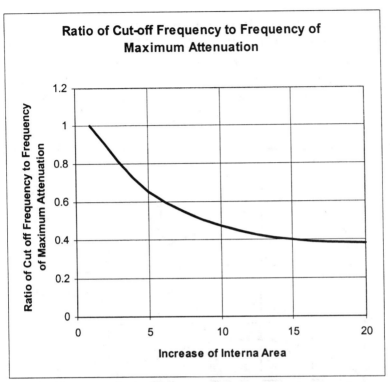

The reactive muffler will transmit frequencies below the cut off frequency. The muffler will also have pass bands at even multiples of the frequency of maximum attenuation. Expansion chambers of different lengths are usually used in series for maximum attenuation.

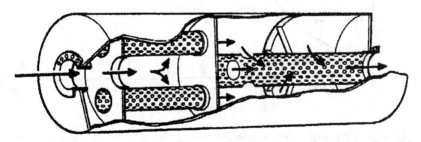

Commercial Muffler with Multiple Expansion Chambers

Dissipative mufflers are not tuned and are typically wide range devices. In their simplest form, they are expansion chambers lined with glass or ceramic fibers. More complex designs split the gas flow into two or more lined tubes or paths. In critical applications, a dissipative muffler is added 10 or more pipe diameters after a reactive muffler.

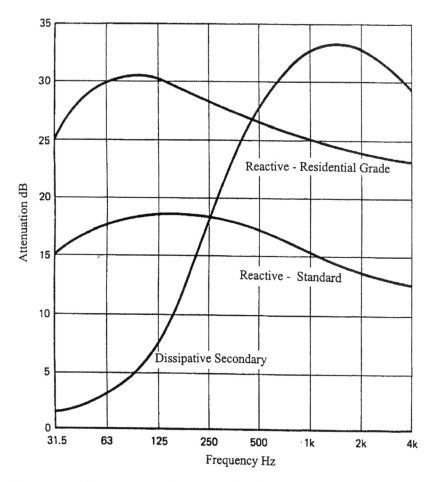

Example: Size an expansion chamber for a 1 ½ in diameter pipe for an expansion ratio of 16.

264

$$\text{Area} = .7854 D^2$$

$$\text{Area}_{\text{pipe}} = .7854 (1.5)^2 = 1.76 \text{ in}^2$$

$$\text{Area}_{\text{Muffler}} = 16 \times 1.76 \text{ in}^2 = 28.27 \text{ in}^2$$

$$\text{Diameter of Muffler} = (\sqrt{\text{area}}) / .7854$$

$$\text{Diameter}_{\text{Muffler}} = (\sqrt{28.27}) / .7854 = 6.77 \text{ inches}$$

PRACTICAL APPLICATION: I have built several mufflers for various odd applications and have generally been pleased with the results. However, for most situations, you can buy an automotive type muffler for $20 that will do a good job. The muffler should be sized for exhaust flow, not by pipe size. If the muffler is too large, the exhaust noise passes through, using only the initial large expansion for attenuation. All mufflers work better with increased back pressure, provided it is within the limits set my the engine manufacturer. When sizing an automotive muffler, look for an auto with the same sized (displacement) engine as your generator.

MATERIAL:

Typically, aluminized steel is used for mufflers and is heat resistant up to 1200° F. Corrosive environments require stainless steel where alloy 409 is commonly used for temperatures up to 1500° F. For extremely corrosive applications, 300 series stainless steel is used.

EXHAUST PIPING:

Exhaust systems should be properly designed to prevent excessive back-pressure, leaks, noise, and condensation from draining back into the engine.

Because each gallon of fuel burned creates about a gallon of water, exhaust piping should be joined to the engine through a T or Y connection. The bottom of the T having a trap and drain for condensation. At least 1-foot of flexible pipe should be added between the engine manifold and exhaust system. A six-inch pipe nipple should be added between the manifold and flex tubing to allow for maintenance and prevent excessive strain on the exhaust manifold casting.

Exhaust pipes passing through combustible walls should be fitted with metal ventilated thimbles as seen on solid fuel burning stoves and appliances. Because the exhaust piping may be 800 to 1000° F, it should be wrapped with ceramic fiber blanket such as Kao-Wool to prevent burns in case of accidental contact.

Both pressure and gas velocity may be measured in inches water column or inches mercury. A manometer is a U shaped tube with water or mercury in the bottom of the U. One end of the tube is connected to the duct in question and the difference in the fluid levels between the two sides is measured with a ruler.

Maximum allowable exhaust back-pressure is specified by the engine manufacturer, however if this is not known, **1.5 inches mercury (.736psi) is typical.**

inches mercury = water column / 13.6
inches water column = 13.6 x inches mercury
1 ounce pressure = 1.73 inches water column = .127inch mercury
1 inch mercury = .491 psi

Exhaust back pressure may be estimated by:

$$\text{Pressure (psi)} = (L \times S \times Q^2) / (5185\ D^5)$$

psi = .0361 x inches water column   L = Length of pipe in feet
Q = Gas flow, CFM   D= Inside Diameter (inches) of pipe
S = Gas Specific Weight   Water column = psi / .0361

$S_{(lb/ft^3)}$ = 39.6 / Exhaust temperature + 460 ° F

Standard elbow (radius = Diameter) L_{Feet} = 2.75D

Long Elbow (radius greater than 1.5D) L_{Feet} = 1.67D

Flex Tubing L = 31 x Connector Length in Feet

Automotive Mufflers - Approximate Exhaust Flow Rate, CFM for a Pressure Drop of 1-inch Mercury or Less

Inlet Pipe Area Sq-Ft	Pipe Diameter	CFM
.0031	3/4	40.5
.0055	1	66
.0085	1-1/4	113
.0123	1-1/2	153
.0014	1.75	211
.0218	2	255
.0341	2-1/2	360
.0491	3	555

Example: Find the back pressure for an exhaust system for a flow rate of 161 cfm at 1215°F. The pipe run is 12 feet, there is 1 elbow and 1 T section and a 1.5 foot section of flex tubing.

Looking at the muffler chart, the proper size is between 1.5-inch inlet and 1.75 inch inlet. I will try the 1.5-inch size first:

The specific weight of the gas S:

$$S= 39.6 / (1215 + 460) = .0236 \text{ lb} / ft^3$$

2   (90° sections) = 2 x 2.75 x 1.5 = 8.25 feet (equivalent)

Flex Tubing = 31 x 1.5 = 46.5 feet (equivalent)

Total equivalent feet of pipe = 12 + 8.25 + 46.5 = 66.75 feet

Pipe Back Pressure = $(66.75 \times .0236 \times 161^2) / (5185 \times 1.5^5)$

Pipe Back pressure = 1.037 inches Mercury

Total pressure with the Muffler = 2.037 inches mercury. This is a little high. You might get away with it if you were doing much part load running but the pipe size should be larger. A quick calculation with 1.75-inch pipe yields:

$$(68.125 \text{ x} .0236 \text{ x } 161^2) / (5185 \text{ x } 1.75^5) = .490\text{-inch mercury}$$

Due to larger muffler's increased capacity, the back-pressure should be lower than 1-inch mercury. Changing to a 1.75-inch diameter pipe and using the larger muffler is a much better choice.

PIPE HANGERS: Steel exhaust piping expands approximately .008-inch per foot per 100°F change in temperature. Assuming a temperature of 1000° and considering the exhaust system above:

$$(1000° / 100) \text{ x } 12 \text{ feet x } .008 = .96\text{-inch}$$

The pipe expands approximately 1-inch. *Exhaust piping should never be rigidly mounted but should have flexible hangers.* Spring type hangers are available however rubber automotive type exhaust hangers are both locally available and inexpensive.

WATER TRAP: Each gallon of fuel burned produces about a gallon of water. A water trap should be included to catch condensation before it can drain back into the engine.

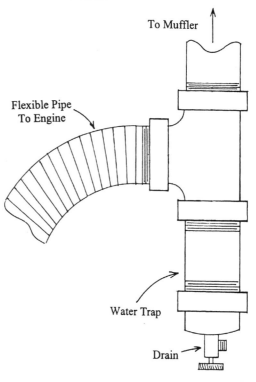

Exhaust Water Trap

Rain Caps should be installed on vertical exhaust stacks. Cutting the pipe off at a 45° angle is usually enough to keep the rain out of a horizontal stack. Do not face the opening into a prevailing wind.

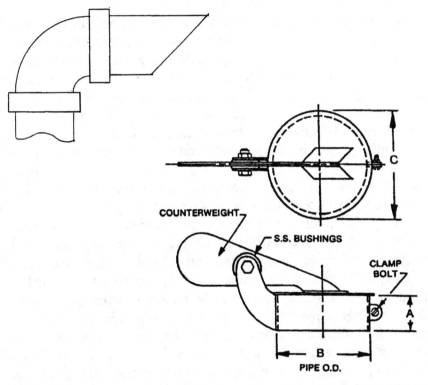

Size	Part No.	A	B		C
			Min	Max	
1	80-1061	1-1/2	1-1/4	1-7/16	1/11/16
1-1/2	80-1062	1-1/2	1-3/4	1-15/16	2/3/16
2	80-1063	1-1/2	2-1/4	2-7/16	2/11/16
2 1/2	80-1065	1-1/2	2-3/4	2-15/16	3/3/16
3	80-1065	1-1/2	3-1/2	3-11/16	3/15/16

Universal Silencer (608) 873- 4272

49. CORROSION:

One fundamental law of nature is that things move towards the lowest possible energy state, for example hot steel cools to room temperature, radioactive materials decay, wood rots, and children would rather sit on the couch than do chores.

Substantial energy input is required to extract pure metals from their ores. There is a tendency of most metals, particularly ferrous metals (iron and steel), to change back to a lower energy state or corrode to a "more stable compound." Usually, these are the same compounds as found in ore from which the metal was extracted.

Corrosion is an electro-chemical process caused by small electric currents between the base metal and the electrolyte. When moisture is present between a metal and some impurity or different metal, a small difference in voltage is set up causing a slight current to flow, oxidizing (corroding) the metal. This is called galvanic corrosion because it is similar to a battery.

Every metal has a tendency to corrode in a particular environment. One method for comparing the tendency of a metal to corrode is by comparing its half-cell voltage to that of hydrogen. Metals that are more reactive than hydrogen are assigned a negative numbers and are said to be **anodic** to hydrogen. Metals that are less reactive than hydrogen are given positive numbers and said to be **cathodic** to hydrogen. The potentials of metals are listed on the next page. The corroding metal will have a greater tendency to give up electrons and will appear to be the more negative metal relative to the other in the oxidation table.

For example: Zinc relative to Iron:

Zn = -.763 volt, Iron = -.440 Volt

-.763 –(-.440) = -.323 volt   Zinc will corrode before the iron, therefore protecting the iron from corrosion.

OXIDATION – CORROSION REACTION

More Cathodic	$Au \rightarrow Au^3 + 3e^-$	**+1.498**
Less tendency	$2H_2O \rightarrow O_2 + 4H^+$	+1.229
to corrode	$Pt \rightarrow Pt^2 + 2e^-$	**+1.200**
	$Ag \rightarrow Ag^+ + e^-$	+0.799
	$2Hg \rightarrow Hg_2^{2+}$	+0.788
	$Fe^{2+} \rightarrow Fe^3 + e^-$	+0.771
	$4(OH)^- \rightarrow O_2 + 2H_2O + 4e^-$	+0.401
	$Cu \rightarrow Cu^{2+} + 2e^-$	+0.337
	$Sn^{2+} \rightarrow Sn^{4+} + 2e^-$	+0.150
	$H_2 \rightarrow 2H^+ + 2e^-$	**+0.000**
	$Pb \rightarrow Pb^{2+} + 2e^-$	-0.126
	$Sn \rightarrow Sn^{2+} + 2e^-$	-0.136
	$Ni \rightarrow Ni^{2+} + 2e^-$	-0.250
	$Co \rightarrow Co^{2+} + 2e^-$	-0.277
	$Cd \rightarrow Cd^{2+} + 2e-$	-0.403
	$Fe \rightarrow Fe^{2+} + 2e^-$	-0.440
	$Cr \rightarrow Cr^{3+} + 3e^-$	-0.774
	$Zn \rightarrow Zn^{2+} + 2e^-$	-0.763
Greater tendency	$Al \rightarrow Al^{3+} + 3e^-$	**-1.662**
to corrode	$Mg \rightarrow Mg^{2+} + 2e^-$	**-2.363**
More Anodic	$Na \rightarrow Na^+ + e^-$	**-2.714**

Tin relative to iron: Sn = -.136, -.136 –(-.440) = +.304 volt

Iron is more negative than tin and will corrode first.

Copper in contact with iron causes rapid corrosion of the iron.

Cu = +.337 +.337 –(.440) = +.777 volt

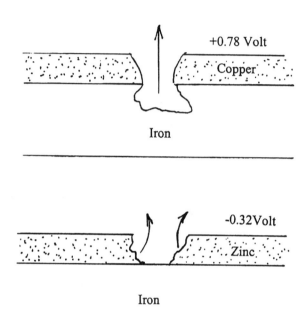

RUSTING OF IRON: If a piece of iron is immersed in oxygenated water (containing air), ferric hydroxide $Fe(OH)_3$ (rust) will form on its surface. If oxygen and water are present, they will react to from hydroxyl ions. $O_2 + 2H_2O + 4e^- \rightarrow 4\ OH$

When combined with iron, the reaction is:

$2Fe + 2H_2O + O_2 \rightarrow 2Fe^{2+} + 4OH^- \rightarrow 2Fe(OH)_2 \downarrow$ precipitate

The precipitate is further oxidized to ferric hydroxide or rust

$$2Fe(OH)_2 + H_2O + \tfrac{1}{2}O_2 \rightarrow 2Fe(OH)_3 \downarrow \text{ Rust precipitate}$$

CORROSION PROTECTION: If current flows from + to -, then a structure is protected if current enters from the electrolyte. For a metal to be protected, electrons must be supplied or it must be made relatively more cathodic. This may be accomplished by adding a sacrificial electrode of a relatively more negative metal or by supplying electrons via a small electric current. Typical metals for sacrificial electrodes relative to iron in order of reactivity* are: magnesium, magnesium alloys, zinc (very low iron zinc is preferred) and aluminum. Note that iron contamination of the zinc reduces effectiveness as an electrode.

*note that this reactivity is from actual corrosion experience. Some metals (aluminum) form films reducing their capacity as a sacrificial electrode.

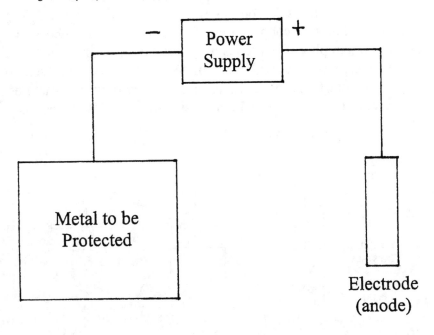

CURRENT REQUIRED FOR PROTECTION: Bare steel tanks with room temperature fresh water require about 3 to 5 milliamps per square foot of surface area, where 1 milliamp = .001amp. Current may be supplied by sacrificial electrodes or impressed by a transformer and rectifier.

Temperature increases the rate of almost all chemical reactions. An increase in temperature causes an exponential rise in the corrosion rate. Hot water tanks over, 140° F usually require 10ma per square foot. While zinc protects steel at room temperatures, it is temperature sensitive and the potential reverses above 140° F. For this reason, magnesium rods are used in hot water heaters.

A structure may also be considered protected if it meets a specified voltage difference between the structure and a test electrode. Steel structures are protected if they are polarized to a potential of -.85 volt verses a copper-copper sulfate electrode. A copper sulfate electrode consists of a copper rod suspended in a saturated solution of copper sulfate that is enclosed in a tube with a porous bottom plug. The porous plug is about 1 inch in diameter. A small sponge soaked with electrolyte or salt water is used to improve contact. Copper sulfate, in contact with steel structures, can form a small galvanic cell, increasing corrosion at that point.

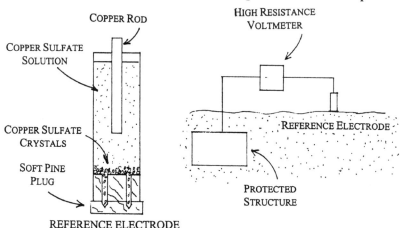

COPPER ROD

HIGH RESISTANCE VOLTMETER

COPPER SULFATE SOLUTION

COPPER SULFATE CRYSTALS

SOFT PINE PLUG

REFERENCE ELECTRODE

REFERENCE ELECTRODE

PROTECTED STRUCTURE

ELECTRODES AND ELECTRODE CONSUMPTION: Electrodes may be categorized as sacrificial or as the impressed current type. However, excluding platinum in impressed current applications, all electrodes are consumed.

Sacrificial electrodes are consumed at the rate of: Zinc-25lbs/amp-year, Magnesium-18lbs/amp-year, Aluminum-tin -20lbs/amp-year.

Impressed current electrodes include: scrap steel- 20 lbs/ampyear, aluminum- 12 lbs/ampyear, stainless steel 1 lb/ampyear and platinum- no consumption.

Resistance of rod type anodes approximately doubles when 80% of the material is gone and they should be replaced.

CORROSION OF CYLINDER LINERS: Corrosion of cylinder liners is a special form of pitting due to the formation and collapse of bubbles in the liquid coolant near the metal surface. It is estimated that rapidly collapsing bubbles produce shock waves with pressures as high as 60,000 psi. This deforms the metal surface, tearing away minute metal particles. It also destroys the protective oxide film. The metal then corrodes and a new film is formed. The process repeats itself eventually forming deep holes. Special "Diesel Rated" coolant reduces this type of corrosion by forming a protective coating over the metal parts.

EROSION OF COPPER TUBING: Copper is susceptible to erosion corrosion, which is an increased rate of deterioration due to the movement of a fluid over the surface. Flow in cool water applications should be limited to 8 feet per second. For hot water applications with temperatures up to 140° F, flow rates should be limited to 5 feet per second. Over 140° F, flow should be limited to 2 feet per second.

STAINLESS STEEL: Chromium is added to steel to make it corrosion resistant. Small amounts of about 5% improve the corrosion resistance but the minimum amount required to make it truly stainless is 12%. Chromium protects the iron surface by forming an oxide layer with prevents the underlying metal from corrosion.

There are two classes of stainless steel, 400 series and the 300 series. The 400 series or ferritic stainless steels are iron-chromium alloys, which contain 12 to 30% chromium. The 300 series contains 16 to 25% chromium and 7 to 20% nickel. The nickel addition raises the cost of the alloy, however it improves the weldability.

Corrosion of carbon containing stainless steels usually occurs at the grain boundaries. Between 900 and 1400° F, chromium combines with the carbon forming chromium carbide. The grain boundary region becomes anodic to the rest of the alloy. If the chromium content is reduced below the critical 12%, the metal is unprotected and corrosion occurs.

Composition of Stainless Alloys					
Type	Cr	Ni	C max	Ti or Mo	Typical Applications
304	19	9	0.08		Chemical and food processing equipment
304L	19	10	0.03		Extra low carbon type of 304 for welding
316	17	12	0.08	Mo 2.5	More corrosion resistant than #304
409	11		0.08	6 x carbon	General purpose- automotive exhaust
442	20		0.2		Furnace parts, combustion chambers
446	25		0.2		high temperature corrosion resistance, sulfur resistance

Corrosion resistance in welded stainless steels may be restored by heat treatment to dissolve the chromium carbides back into solution. Heat treatment consists of heating and holding the metal at approximately 1985° F, finally quenching. If the metal must operate in the 900 to 1400° F range, it is typically stabilized by adding titanium or columbium. These additions are very strong carbide formers and leave almost no free carbon to combine with the chromium. Molybdenum increases the resistance to pitting.

Stainless steel in contact with other metals may have a significant effect on the extent of damage due to galvanic corrosion. A large cathode and a small anode increases the current density at the anode, therefore carbons steel bolts in a stainless steel plate will corrode rapidly. However a large carbon steel plate with stainless steel bolts with see only slightly increased corrosion

Example: Estimate the required impressed current for a 500-gallon mild steel water tank if the temperature of the water is 200° F. The tank is a salvaged propane tank with hemispherical heads. The diameter of the tank is 41 inches and the total tank length is 104 inches.

Find the surface area:

Subtracting the hemispherical ends, the length of the straight section of the tank is: $104 - 41 = 63$ inches

The surface area of the straight section of the tank = πDL

$\pi = 3.14$, D = 41 inches, L = 63 inches

$\qquad 3.14 \times 41 \times 63 = 8115$ inches^2

The surface area of a sphere = πD^2 = $3.14 \times 41^2 = 5281$ inches^2

Total surface area = $8115 + 5281 = 13,396$ inches^2

1 foot $^2 = 144$ inches ^{2}

Surface area = $13,396 / 144 = 93$ ft^2

Current for 140°+ hot water = 10ma / ft^2

93 ft^2 x 10 amp x .001 = .93 amp

The required protective current = .93 amp

A small 1-amp battery charger with a 3-foot long ¾-inch diameter steel rod as an electrode will be used to protect the tank from corrosion.

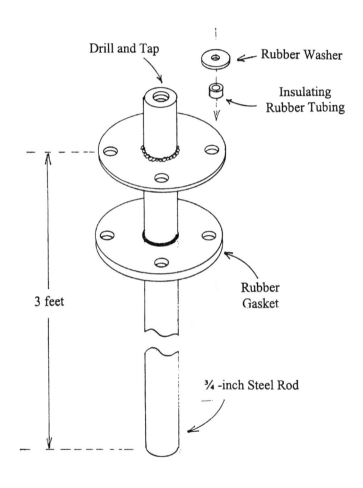

Electrode Detail

50. FUEL TANKS:

If you are installing a large tank, there are guidelines to make the installation more user friendly and less hazardous. This is not an exhaustive presentation of all the applicable principles and codes but a brief list of suggestions. There are several documents available for review such as PEI 100 Recommended Practices for Installation of Underground Liquid Storage Systems, by the Petroleum Equipment Institute; also NFPA 30 Flammable and Combustible Liquids Code, by the National Fire Protection Association and American Petroleum Institute's API 1615 Installation of Underground Petroleum Storage Systems. Equivalent Publications are available for above ground storage tanks or AST's. The local fire marshal should be consulted for local codes. Zoning laws may regulate the installation, size, and locations of a tank. If your installation requires inspection, you should purchase a UL listed tank with the listing on the nameplate. UL specifies the steel gage thickness, weld profiles, fittings, bulkheads and leakage tests.

Tank Size:

Emergency power systems may use small tanks and many power systems are rated by the number of hours the set will run on a particular tank. My original standby power plant has a gasoline tank fabricated from an old propane tank. It holds about 8 gallons and the set runs from 6 ½ to 10 ½ hours, depending upon the load.

If you are running your generator as the primary source of power, probably the best size fuel tank for diesel or propane is 500 gallons. Fuel is much cheaper is 500-gallon lots and as long as a tank is moveable on your property as opposed to permanently installed, you are not subject to much of the permitting and many of the regulations. Skid mounted diesel tanks are typically deemed temporary installations. Usually, you are able to have more than one temporary tank without falling under excessive regulation.

A short-run tank gasoline tank fabricated from a discarded propane tank

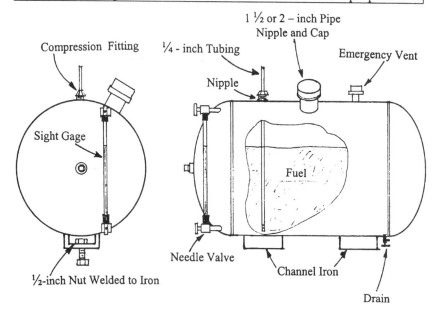

Compression Fitting

¼ - inch Tubing

1 ½ or 2 – inch Pipe Nipple and Cap

Emergency Vent

Nipple

Sight Gage

Fuel

Needle Valve

Channel Iron

Drain

½-inch Nut Welded to Iron

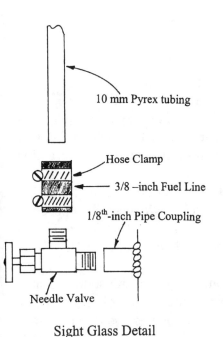

10 mm Pyrex tubing

Hose Clamp

3/8 –inch Fuel Line

1/8th-inch Pipe Coupling

Needle Valve

Sight Glass Detail

I have made several small tanks from salvaged propane cylinders as seen on the previous page. The tank easily bolts to a machine frame using the nuts welded inside the channel iron at the bottom of the tank. Two or three washers are placed underneath the mount farthest from the drain. This angles the tank so that any water will accumulate at the drain. My original tank does not have a sight gage. The fuel level is easily checked with a ¼-inch dowel through the filler neck. The sight gage has needle valves so that the flow may be turned off in the event of a fuel leak. Note that these fabricated tanks do not necessarily meet any particular safety standard. The sight glass should be protected.

TANK INSTALLATION:

Tank Foundations:

A typical 500-gallon tank weighs a little 4500 to 5000-pounds when full. While most soil will support a skid-mounted tank, the consequences of tank settlement over time can be severe. You should avoid locating a tank over an area that has been previously excavated, especially if it has not been properly compacted. Probably the greatest problem caused by tank settlement is the stress placed on the piping attached to the tank. A flexible connection in the fuel line, a loop in the fuel line, and a reinforced concrete pad reduce these problems for little cost.

Dikes:

Oil spills typically result from leaking tanks, piping or overfilling. Dykes are built around tanks to contain spills for fire fighting and environmental protection. Dykes may be made of soil or concrete. While soil dykes are not liquid tight and offer little environmental protection, they do prevent liquid from draining towards buildings and property lines and are of value as far as fire fighting is concerned. Concrete dykes may crack over time. The bottom and sides are usually poured as a single piece as opposed to block construction. Containment is considered leak-proof if it contains the spill for 3 days without contaminating the surroundings.

The floor of the dyked area should slope towards a drain that is controlled outside of the dyke. Dyked areas can collect rainwater and often have a rainshield or a roof. Rainwater should be drained from the dyke for a number of reasons including mosquito control and to prevent creating hazardous waste if there is an oil spill. Most dykes are designed to hold 110% of the tank's capacity while a few

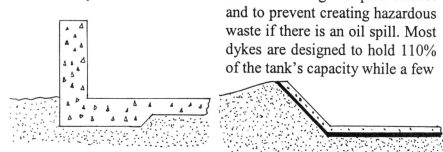

Monolithic Dikes

states require 125%. The sides of a dyke should not be too close to the sides of the tank. A leak in the side of a tank may jet several feet completely missing the dyke.

Dyked tanks are subjected to a buoyancy test. The dyke is filled with water to its maximum capacity while the tank remains empty. The tank should not float away. *See example.

Protecting the Tank from Damage:

Besides fires and spills, there are two other types of damage a tank may be protected from, vehicle impact and bullet proofing.

Vehicle Impact:

If a tank is located in an area where it could be struck by a vehicle, it may be required to have impact protection. This is typically concrete filled pipe that is attached to or set into a concrete foundation. Inspection of a few gasoline pumps in your area should reveal this type of protection.

Bullet Proofing:

If the tank is located in an area where stray or intentional bullets may strike it, fire officials may require it to be "projectile resistant." Test tanks are subjected to 5 gunshots from a 30-caliber rifle located 100 feet from the tank. 150-grain military ball ammunition is used in the test. A tank may be listed as projectile resistant if it passes a hydrostatic test after the ballistics test. This does not mean that someone will test your tank by shooting at it. A similar tank has been tested and the results are on file.

Tanks that are not listed as projectile resistant may be protected by bullet resistant construction which includes 8-inch thick solid concrete, 8-inch concrete block walls that have been filled with a sand and concrete mixture, or 5/8-inch thick steel plate.

Access:

Fire codes have requirements for fire department access to a facility, which usually includes both adequate roadway and vertical clearance for fire trucks. Similar access and clearances are required for a fuel truck to deliver fuel. Although hoses may be 100 feet long, a tank is best located within 50 to 75 feet of an accessible drive. A supply of water may also be required.

Water Supply:

Water has a high heat of vaporization, approximately 8,100 Btu per gallon, making it very effective fire fighting agent. A liquid filled tank will absorb heat from a pool of burning oil at a rate of about 20,000 Btu / ft^2 per hour. Applying water to the tank will reduce this to about 6000 Btu / ft^2 per hour. The minimum water flow rate is .25 to .5 gallon per minute per ft^2 wetted area. The wetted area of the tank is the actual inside surface area of the tank that is covered by liquid.

Tank Vents:

Above ground fuel storage tanks have vents and emergency vents. Normal vents protect the tank from collapsing from a vacuum as fuel is removed and emergency vents prevent an explosion during a fire. Normal vents for tanks 2500 gallons or less are made of 1 ¼-inch diameter pipe. For Class I, II, and III liquids they are typically raised 12 feet above grade to prevent accidental ignition of the vapors. Emergency vents are sized in the table "Emergency Venting Capacity." The fire codes and NFPA 30 requirements for sizing emergency vents have a 4:1 safety factor that is included in the table.

Correctly sizing an above-ground tank vent has three steps:

1. Calculation of the tank's wetted area.

2. Determining the emergency vent flow rate. Assume 1055 ft^3 of vapor are generated per ft^2 of wetted area per hour.

3. Ensuring the emergency vent will adequately relieve the vapor flow. Assume a vent will safely flow 6000 to 7500 ft^3 per $inch^2$ per hour.

When a tank is involved in a fire, the temperature rises until boiling occurs. The emergency vent must be large enough to prevent pressure rise in the tank.

Horizontal Tanks: NFPA requires that 75% of the tank surface area be considered when sizing the emergency vent. This is based on the assumption that 75% of the area will be heated by optically thick flames which are so intense that energy absorbed by the tank can not be radiated back to the atmosphere. These flames are 3 to 6 feet high in hydrocarbon pools.

Classification of Flammable and Combustible Liquids

Classification	Flash Point Temperature
1 – C Flammable	Greater than 73° Less than 100°F
Class II Combustible	Greater than 100 °F Less than 140 °F
Class III-A Combustible	Greater than 140 °F Less than 200 °F
Class III-B Combustible	Greater than 200 °F

Emergency Venting Capacity for Steel Tanks:

Wetted Area ft^2	Venting Capacity ft^3 / hr	Pipe Size
20	21,100	2
30	31,600	2
40	42,100	3
50	52,700	3
60	63,200	3 ½
70	73,700	4
80	84,200	4

90	94,800	4
100	105,000	4
120	126,000	5

Vertical Tanks: The lowest 30 feet of a vertical tank are used to size emergency vents for such tanks. Large hydrocarbon pool fires are about 30 feet tall.

Example: Size an emergency vent for a small horizontal cylindrical fuel tank that is 12 inches in diameter and 18 inches long.

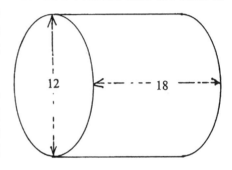

The area of the 2 tank ends = 2 (.7854 D^2)

The area of the tank ends = 2 (.7854 x 12^2) = 226.2 inches^2

The area of the sides = πD x L, π = 3.14

The area of the sides = (3.14 x 12 x 18) = 678.2 inches^2

The total area = 226.2 + 678.2 = 904 inches^2

The area in ft^2 = (904 inches^2) / (144 inches^2 per foot^2) = 6.28 ft^2

Because it is a horizontal tank, 75% of the area is used to size the emergency vent. .75 x 6.28 = 4.7 ft^2

The vapor flow rate = 1055 ft^3 per ft^2 per hour

The vapor flow rate 4.7 ft^2 x 1055 ft^3 / ft^2 = 4959 ft^3 per hour

Inside area of 1-inch pipe (from chart) = .864 inch^2

4959 ft^3 per hour / .864 inch^2 = 5739 ft^3 per hour per inch^2

A 1-inch diameter vent flows 5739 ft^3 / hour / inch^2, which is less than the maximum of 7500 ft^3 / hour / inch^2. A 1-inch vent will be used.

Opening Pressure:
Emergency vents should open below 2.5 psig.
Find the maximum weight of the vent top from the preceding example: (.864-inch^2) x (2.5 psi) = 2.16 pounds

The vent lid should weigh less than 2.16 pounds and assuming some friction, the lid should weigh less than 2 pounds.

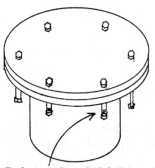

Bolts not threaded full length
Bolts do not tighten against vent

Internal Area of Schedule Steel 40 Pipe:

Pipe Size	Area inches^2	Pipe Size	Area inches^2
1	.864	2	3.35
1.24	1.5	3	7.07
1.5	2.04	4	12.6

Leak Testing: Horizontal tanks are pressurized to 5 psig and painted with a solution of soapy water. Bubbles indicate the presence of a leak that should be repaired before the tank is put into service. Large vertical tanks are tested at 2 ½ psig. Many fire or building code inspectors require that a representative be present to witness a leak test. You should perform the test in advance to be sure your tank will pass.

Adequate Separation Distance (ASD): Regulations require an adequate separation distance between a tank and other structures. ASD is the actual distance beyond which an explosion or combustion is not likely to cause structures or individuals exposure to thermal radiation (heat) in excess of the safety standards."

Human exposure to heat of 1500 Btu / ft^2 / hr causes intolerable pain within 15 seconds. Longer exposure causes blistering, permanent skin damage and possible death. Children and the elderly may not be able to move away from a fire quickly enough; therefore the level for unprotected areas where people may congregate is 450 Btu / ft ^{2} / hr. Exposure to this level for a prolonged time is similar to a severe sunburn.

Wooden buildings and trees will ignite within about 15 minutes when exposed to 10,000 Btu / ft^2 / hr. Therefore the thermal radiation level at a structure shall not exceed 10,000 Btu / ft^2 / hr.

ASD is measured from the center of the tank and is based on the area of the dyke around a tank. The fire width is estimated by:

$$\text{Fire Width} = \sqrt{\text{Dike Area}}$$

Acceptable Separation Distance:

Dyke Area	Fire Width	ASD Buildings 10,000 Btu / ft^2 / hr	ASD Persons 450 Btu / ft^2 / hr
100	10	10	60
200	14	13	80
300	17	15	90
400	20	17	107
500	22	18	115
600	24	21	122
700	26	22	130
800	28	24	140

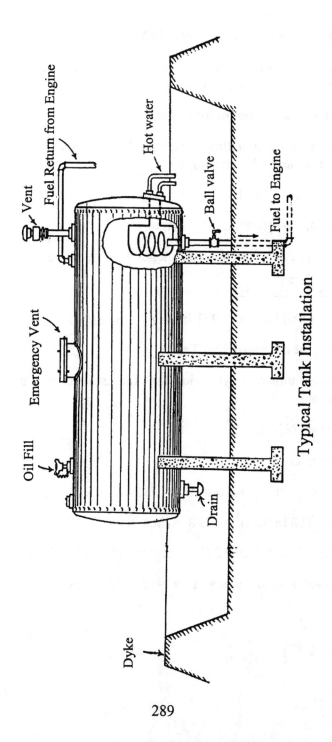

Vent

Fuel Return from Engine

Hot water

Ball valve

Fuel to Engine

Emergency Vent

Oil Fill

Drain

Dyke

Typical Tank Installation

289

ESTIMATING TANK CAPACITY:

Calculation of tank capacity is a straight froward process involving calculation of volume and conversion of units.

Capacity of a Rectangular Tank:

Find the volume, in gallons, of a tank that is 30 inches long, 18 inches wide and 12 inches high.

Volume = L x W x H

L = length, W = width, H = height

V = 30 inches x 18 inches x 12 inches = 6480 inches ^{3}

$\boxed{1 \text{ gallon} = 231 \text{ inches}^3}$

Answer: 6480 inches ^{3} / 231 inches^3 = *28 gallons*

Capacity of a Cylindrical Tank:

Find the volume of a cylindrical tank that is 6 feet in diameter and 10 feet long.

Volume = .7854 D^2 L

D = diameter   L = length

V = .7854 ((6 feet)^2 x 10 feet) = .7854 x 36 x 10 = 282.7 foot^3

1 foot^3 = (12 inches)^3 = 1728 inches ^{3}

1 foot^3 = 1728 inches^3 / 231 inches^3 = 7.48 gallons/foot^3

282.7 foot^3 x 7.48 gallons / foot^3 = *2115 gallons*

				Capacity of Flat Ended Cylindrical Tanks Per Inch of Length			
Diameter inches	Gallons	Diameter inches	Gallons	Diameter inches	Gallons	Diameter inches	Gallons
10	0.340	27	2.479	44	6.582	61	12.651
11	0.411	28	2.666	45	6.885	62	13.070
12	0.490	29	2.859	46	7.194	63	13.495
13	0.575	30	3.060	47	7.511	64	13.926
14	0.666	31	3.267	48	7.834	65	14.365
15	0.765	32	3.482	49	8.163	66	14.810
16	0.870	33	3.703	50	8.500	67	15.263
17	0.983	34	3.930	51	8.843	68	15.722
18	1.102	35	4.165	52	9.194	69	16.187
19	1.227	36	4.406	53	9.551	70	16.660
20	1.360	37	4.655	54	9.914	71	17.139
21	1.499	38	4.910	55	10.285	72	17.626
22	1.646	39	5.171	56	10.662	75	19.125
23	1.799	40	5.440	57	11.047	78	20.686
24	1.958	41	5.715	58	11.438	84	23.990
25	2.125	42	5.998	59	11.835	96	31.334
26	2.298	43	6.287	60	12.240	120	48.960

This problem may also be solved by using the table above.

6 feet x 12-inches / foot = 72-inches

Gallons per inch of length of a 72-inch diameter tank = 17.626

10 feet x 12 inches / foot =120-inches

120-inches x 17.626 gallons / inch = *2115 gallons*

Fuel in a Partially Filled Cylindrical Tank:

How many gallons are in the cylindrical tank from the previous example if the fuel is 2.5 feet deep?

1. Find the percentage of depth: 2.5 feet / 6 feet = .416

2. Convert depth to volume using the chart below.

Estimating, .416 is approximately 39% of the volume

Fractional Capacity of Partially Filled Horizontal Cylindrical Tanks			
Depth	Volume	Depth	Volume
0.05	0.0187	0.55	0.5636
0.10	0.0520	0.60	0.6264
0.15	0.0941	0.65	0.6881
0.20	0.1424	0.70	0.7477
0.25	0.1955	0.75	0.8045
0.30	0.2523	0.80	0.8576
0.35	0.3119	0.85	0.9059
0.40	0.3736	0.90	0.9480
0.45	0.4364	0.95	0.9813
0.50	0.5000	1.00	1.0000

Note that dished (convex) tank heads holds 67% as much as the straight section of tank.

3. Multiply the capacity of the tank by the percent volume.

2115 gallons x .39 = 825 gallons

The tank contains 825 gallons.

Tank with Hemispherical Ends:

Find the capacity of a horizontal tank with hemispherical ends. The tank is 16 feet long and 6 feet in diameter.

The tank is calculated as a sphere plus a cylinder.

The tank is 6 feet in diameter, therefore the 2 hemispherical ends form a sphere 6 feet in diameter.

The volume of a sphere = .5236D^3

The volume of a 6 foot diameter sphere = .5236 (6ft)^3 = 113 ft^3

The volume of the 6 foot diameter by 10 foot long tank from the previous problem = 283 foot^3

283 ft^3 + 113 ft^3 = 396 ft^3

396 ft^3 x 7.48 gal./ ft^3 = 2962 gallons

The tank capacity is 2962 gallons.

This problem may also be solved using the fact that hemispherical heads hold 67% as much as the straight cylindrical section.

(.67 x 72-inches + 120-inches) x 17.626 gallons / inch = 2965 gallons. Note that the 3 gallon difference is due to rounding .6666 up to .67

293

Tank Capacity, Leaks, Buoyancy and Dyke Sizing:

The maximum distance of a jet of fluid from a hole in a tank is relative to the height of fluid in the tank and is located at the middle of fluid height in the tank. The horizontal distance of a tank leak from the hole to the level of the tank base is given by :

$$\textbf{Distance} = 2 \sqrt{h(H-h)}$$

H = total height of liquid,   h = height from tank base to hole

Example: Find the horizontal distance of the jet from a hole located in the middle of a 6 - foot diameter horizontal tank.

Distance = $2 \sqrt{3(6-3)}$ = 6 feet

Dyke design:

Design a dyke for a horizontal cylindrical fuel tank that is 5 feet in diameter and 10 feet long: Assume that the tank is not elevated and that there is 16-inch high dyke wall.

The length of the horizontal jet is relative to both the *velocity* of the fluid leaving the hole and the *time* required for the fluid to reach the ground. The height of the hole above the ground dictates the time.

The velocity of the jet at the hole in feet per second is: $V = 8\sqrt{h}$

V= velocity of discharge   h= height of fluid above the hole

Time required for the leak to reach the ground is:

$$T = \sqrt{.062\, y}$$

T = time in seconds y = height of the jet above the ground

Assuming that the hole is located at the middle of the tank, the velocity of the jet is:

$V = 8\sqrt{2.5}$ $V = 12.65$ feet per second

The time required for the jet to reach a height 16 inches above the ground is:

$T = \sqrt{.062\ \ (2.5\ \text{feet} - 1.33\ \text{feet})}$ note (16 in / 12 in. per ft) = 1.33 feet

$T = .27$ second

The horizontal distance the jet flows is:

Distance = TV = .27 second x 12.65 feet / second = 3.42 feet

Answer: A 16-inch high dyke wall must be at least 3.42 feet from the edge of the tank.

Now determine if the dyke is deep enough to contain 125% of the tank contents.

The volume of the cylindrical tank is:

Volume = $.7854D^2\ L$ D = diameter,   L = Length

Volume = $.7854\ (5\text{ft})^2$ x 10 ft = 196 ft^3

Required dyke volume = 1.25 x 196 ft^3 = 245 ft^3

The volume of the dyke is: 11.84 ft x 16.84 ft x 1.33 ft = 265 ft^3

The dyke is slightly larger than required by regulation.

Find the capacity of the tank in gallons:

There are 7.48 gallons per foot ^{3}

The capacity of the tank is 196 ft^3 x 7.48 gal/ ft ^{3} = 1466 gallons

Estimate the Tank Buoyancy or "will it float when the dyke is full?"

Find the amount of steel in the cylindrical tank:

Surface Area of the tank = $1.57 D^2 + \pi D \times L$

Surface Area of the Tank = $1.57 (5^2) + 3.14 \times 5 \times 10 = 196$ ft^2

Assuming the tank is .375 inches thick,

196 ft^2 x (.375 inches / 12 inches / foot) = 6.125 ft^3

Specific Gravity of iron = 7.75

6.125 ft^3 x 7.75 = 47.468 ft^3

Find the depth of the tank that is under water:

1.33 feet / 5 feet = .266

Using the table fraction tank capacity approximately 21% of the tank capacity is under water.   The specific gravity of water is 1

196 ft^3 x .21 = 41.6ft^3

At 47.468 the weight of the iron in the tank is greater than the water displaced by the tank at 41.6.

The empty tank should not float if the dyke is filled with water.

Note: A more accurate calculation of tank weight would be to take the difference of outside and inside volumes of the tank multiplied by the steel weight per unit volume. However the difference from the above method is usually 1% or less.

51. SALVAGED TANKS:

Used or "scrap" propane tanks are available for little cost in most parts of the country. Because these tanks are pressure vessels, They are taken out of service when they develop rust pits. Such tanks are typically rated at 250 psi at 600° F. Wall thickness is usually .375 for 500 and 1000 gallon tanks, .25 for 250 gallon tanks and .1875 for 120 gallon tanks. In comparison the typical wall thickness for a fabricated 500-gallon oil tanks is .070-inch and it is rated at 15-psi.

The quality of scrap tanks varies widely from terrible to almost new. Rust pits typically occur on the bottom of the tank between the legs. I have purchased scrap tanks with only a single dime sized rust pit while others in the same yard were badly rusted. Each yard has different fees for their scrap tanks. One yard gives the tank away for free but asks for a $20 loading fee, while others yards load it for free but may charge as much as $100 for the tank. Since the supply of tanks is variable, it is best to seek out your suppliers and call then on a regular basis. In a few months, you should have a few good scrap tanks to work with.

To get the tank back in as new a condition as possible, I strip the paint from it using a liquid stripper and follow this by grinding out and welding the rust pits, grinding the welds back to the original tank surface. Before welding or cutting, the tank must be ventilated. Propane is heavier than air and pools in the bottom of the tank. Ideally, the tank would be ventilated with the openings at the bottom. The object of the ventilation process is to get the percentage of gas below the **Limits of inflammability.** Not all mixtures of fuel and air will support combustion. The limits of which lean and rich mixtures that do support combustion are called the limits of inflammability. The limits of inflammability for oil vapors are from 1.4% to 7.6%. The limits for propane are from 2.35% to 9.5%.

While I often flip the tank over on its side, put an air hose into one of the openings and let it blow air into the tank for a few hours, however, there is a purging procedure for propane tanks. It involves filling the tank with air to 15 psi, then opening the tank and letting the air escape. This procedure is repeated 6 times. The percentage of gas and air remaining in the tank is seen below:

Purging of LP Gas Containers

# Purging	% Propane Remaining	% Air Remaining
1	50	50
2	25	75
3	12.5	87.5
4	6.25	93.75
5	3.13	96.87
6	1.56	98.44
7	0.78	99.22

After purging, there will be a stinky oil "ethyl mercaptan" left in the bottom of the tank. This oil is used to odorize the gas and gives it that familiar "rotten cabbage smell." Because this oil does burn, however poorly, I will keep an air hose in the tank to ventilate the vapors when cutting and welding.

I first cut the opening for the safety vent. (See the section on proper sizing of safety vents.) I also cut a hole and add a 2-inch pipe coupling so that I may attach a pipe nipple and cam-lok connector for filling. These connectors are used by oil service trucks and are both readily available and very inexpensive. Typically, I pay about $4.00 per half of the aluminum connector.

The difference between an Alcohol and a Mercaptan is that sulfur is substituted for oxygen. Ethyl Alcohol = CH_3CH_2OH Ethyl Mercaptan = CH_3CH_2SH

Male and Female Cam-Lok Connectors with Cover, Plug and Safety Chain. These are threaded 2-inch NPT

These inexpensive Cam-Lok connectors are readily available, in many sizes from ¾ inch to 4-inch, from hydraulic hose shops.

I cut a hole at one end of the bottom of the tank and add an elbow with a valve to make the tank drain. Finally, I rinse the tank with a few gallons of diesel or mineral spirits and paint the outside with a good rust resistant enamel.

One thing to note, I only use discarded *propane* tanks because I can ventilate them. NEVER cut or weld on an old oil tank. They are almost impossible to properly ventilate. The heavier the oil, the more difficult it is to ventilate because the oil stays in cracks and crevices only to flow out when the tank become hot from cutting or welding. Never cut or weld on a tank that has been rinsed with oil.

52. PROPANE VAPORIZATION CAPACITY:

In gasoline engines, gasoline is vaporized in the engine; however in gas engines the gas must be vaporized outside of the engine. High power outputs require high propane vaporization rates. Vaporization rate, temperature and tank pressure are all related properties. Every liquid has a **vapor pressure** that increases as the temperature increases. If the liquid is enclosed in a container, the pressure may be read from a pressure gage. Steam engines work on the vapor pressure of water. Water in the boiler is heated until it reaches 212° F at which point its vapor pressure increases until it begins to boil. As more heat is applied, the pressure increases.

Heat is added to vaporize the liquid propane in a tank. Often the heat of the air or surroundings is sufficient to vaporize the liquid

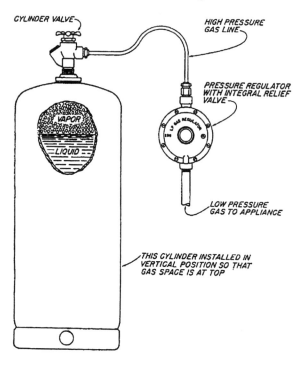

propane. This heat is transferred from the air through the metal tank to the liquid. The surface area of the tank that is covered by liquid propane is called the *wetted surface*. The larger the wetted surface area of the tank and the greater the amount of liquid in the tank increase both the tank's ability to absorb heat and the vaporization rate. Engines will not operate properly without sufficient vaporization capacity. The worst conditions for vaporization are when the tank has a small amount of propane in it and the temperatures are low. A common complaint found when using undersized tanks is "bad gas at the bottom of the tank." The real problem being there is not enough heat flowing into the tank to vaporize the gas. The heat of vaporization for propane at atmospheric pressure is 183 Btu per pound or 774 Btu per gallon. At higher pressures, the heat of vaporization increases.

Maximum Continuous Propane Vaporization in Pounds per Hour per Square Foot of Wetted Area:

Temperature °F	Pounds / Foot^2
-40	.040
-30	.139
-20	.238
-10	.341
0	.440
10	.549
20	.659
30	.733
40	.842
50	.952
60	1.025
70	1.135

Commercial propane is 96% propane, however some propane or LP gas is actually a mixture of propane and butane (the heating value of each is almost identical). At atmospheric pressure, propane boils at - 43.7° F while butane has a boiling point of 31.1° F. The pressure in a tank depends on the temperature and the mixture of gases in the tank.

Vapor Pressures of Propane and Butane

Temperature ° F	Propane psig	Butane psig
-20	10.7	NA
-10	16.7	NA
0	23.5	NA
10	31.3	NA
20	40.8	NA
30	51.6	NA
40	63.3	3.0
50	77.1	6.9
60	92.4	11.6
70	109.3	16.9
80	128.1	22.9
90	149.0	29.8
100	172.0	37.5
110	197.0	46.1
120	225.0	56.1
130	257.3	66.7
140	290.3	77.9

At 80°F, propane has a vapor pressure of 128 psig, while butane has a vapor pressure of 22.9 psig. The pressure in the tank of a gas mixture would be between these two pressures, depending upon the percent composition of the gases. The proportions of the mixture may be identified by its vapor pressure. LP gas products are commonly identified by the vapor pressure at 100° F.

There are two agencies that specify tanks, the Department of Transportation and the ASME or American Society of Mechanical Engineers. Vaporization capacity of two types of standard tanks are shown below:

100 POUND CYLINDERS:

To estimate the number of 100 pound cylinders required for a certain continuous vaporization rate, assume the cylinder is roughly 40% full and the lowest temperature to be 20 degrees below the surrounding air temperature. See the vaporization chart of 100-pound cylinders for the proper figure. Then use the equation shown below:

Number of cylinders = Total BTU / vaporization rate

Calculate the number of cylinders for 300,000 Btu per hour at 60 degrees: Assume 40 degrees F cylinder temperature (60 −20 = 40)

Number of cylinders = 300,000 Btu / 105,000 Btu

Number of cylinders = 2.857 or 3 cylinders

The tanks are connected in parallel to a manifold in order to meet the required vaporization rate. Gas manifolds are available from your gas dealer or should be constructed of schedule 80 pipe and 250 psi fittings.

Some high bulk applications, or those using high percentages of butane, require liquid to be drawn from the tanks and vaporized in a LP gas vaporizer Small vaporizers are common on fork lift trucks where the gas is heated and vaporized using the hot water from the engine cooling system. A commonly used vaporizer is direct fired using gas-burners. Such UL listed vaporizers are available; however, they are very expensive.

303

53. PROPERTIES OF FUEL OILS AND GENERAL SPECIFICATIONS:

Diesel engines are able to burn a wide variety of fuels including mineral oils, animal and vegetable oils. The properties or characteristics of the fuel have considerable influence on the performance and reliability of a diesel engine. While laboratory tests may give some indication of the fuel's performance, actual fuel performance is sensitive to both molecular arrangement and size. Engine tests should be made to evaluate the suitability of any fuel for a particular service.

The principle fuel properties affecting service are:

Ignition quality	Cloud point
Carbon residue	Pour point
Sulfur content	Viscosity
Ash content	Flash point
Water	Heating Value
Sediment	Specific Gravity

Ignition Quality: Fuel oils do not ignite immediately upon being injected into the combustion chamber. The interval of time between the beginning of injection and ignition is called ignition lag and varies widely for different fuels when injected into the same combustion chamber. The shorter the ignition delay, the better ignition quality of the fuel. The longer the delay, the more pronounced is the "Diesel Knock." This is due to the large quantity of fuel that is injected before ignition. When ignition occurs, the flame spreads rapidly through the fuel already in the combustion chamber causing a rapid rise in pressure creating an audible knock.

Cetane ($C_{16}H_{34}$) is a hydrocarbon with a very high and constant ignition quality. If a diesel fuel has the ignition quality of a mixture containing 40% cetane and alpha-methylnapthalene ($C_{11}H_{10}$) with a low ignition quality, then the cetane number of the fuel would be 40.

The cetane number of a fuel is determined by testing it in a standard test engine and varying the compression ratio until the ignition lag is 13 degrees with all other operating conditions (speed, temperature, timing) being constant. The cetane number is the percentage of cetane in a blend of alpha-methylnapthalene that would have the same ignition qualities. High-speed engines require a higher cetane number. Larger, slower diesels with large bores and strokes have more time for ignition and can use fuels with lower cetane numbers.

Ignition accelerators are substances that can be added to reduce the ignition delay period. The most effective, among others are amyl nitrate, ethyl nitrate, and ethyl nitrite. The addition of 1% of ethyl or amyl nitrate raises the cetane value of the fuel. A 5% addition may reduce the ignition temperature by $110°$ F and increase the maximum cylinder pressure by 120 psi.

Viscosity is the opposite of fluidity. It is a measure of a fluid's resistance to flow or shear. Higher viscosity oils do not flow easily, where low viscosity oil flow readily. This is seen in cold weather as gasoline with a low viscosity flows readily while 40-wt motor oil, with a higher viscosity, is quite thick in comparison.

As the temperature of oil is increased, its viscosity is reduced and it flows more readily. A general rule is that an increase of $35°$ F. will reduce the viscosity by one half.

Viscosity also affects injection characteristics. High viscosity fuels tend to give a coarse, penetrating spray rather than a finely atomized one. Increasing viscosity gives increasing exhaust smoke. Higher viscosity fuels may be used if they are preheated.

Low viscosity was believed to have an adverse affect on pump and injector wear, however compared to wear caused by dirt it is minute. Low viscosity fuels may result in increased pump and

injector leakage. At a constant pump setting, this leakage reduces
the volume of fuel delivered.

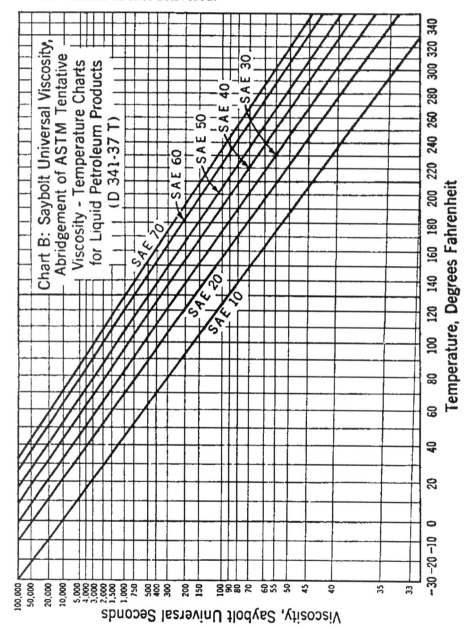

Specific Gravity: The specific gravity of a fuel has a bearing on the injection properties affecting both the depth of penetration and spray cone angle. A lighter fuel will have a smaller depth of penetration and a larger cone angle. The higher the specific gravity, the greater the carbon content and heating value by volume. However, lighter fuels have a greater percentage of hydrogen and a higher heating value per pound.

Pour and Cloud Point: The pour point of a fuel is the lowest temperature at which the oil will flow of its own accord. This determines if a fuel may be pumped at low temperatures. The cloud point is the temperature at which it becomes cloudy due to the formation of wax crystals. This is associated with the clogging of fuel filters and lines.

Flash point: As the temperature of fuel oil increases, vapor is given off and collects at the surface. When the temperature rises to a point where the vapors ignite when exposed to an open flame, it has reached its flash point. The flash point is the highest temperature at which the oil can be stored without being explosively dangerous. Flash point has no correlation to the ignition quality of the fuel or engine performance but is of importance in connection with safety and legal requirements. Gasoline has a low flash point but is a poor diesel fuel. The typical flash points of common oils are listed below:

Diesel	175 to 220°F
Fuel oil	170 to 280°F
Lubricating oils	400 + °F

Fire point. The temperature at which an oil gives off enough vapor to support sustained combustion is called the fire point. The fire point is usually about 50 degrees higher than the flash point.

Carbon Residue: After all the volatile matter in a sample has been evaporated off by heating in a closed container, carbon residue remains. This is a measure of the heavy components that will remain in the engine and form coke.

Sulfur: Sulfur may be present in many forms including hydrogen sulfide, which is corrosive. Burning high sulfur fuels forms sulfur dioxide and trioxide, which form acids in the presence of water. High sulfur fuels require more frequent engine oil changes and or additives to reduce the corrosive effects. Sulfur has a lubricating effect on injector pumps. The sulfur content of fuels for high speed engines may be as high as 1%. Low speed engine may tolerate as much as 3%. Air quality standards have dramatically reduced the amount of sulfur in commercial fuels thereby reducing the lubricity of the fuel. Reduced lubricity was associated with early injection pump failures. Recently, the lubricity issues have been addressed. High sulfur content has also been associated with higher particulate matter in the exhaust.
Note that current US sulfur standards allow .2% by weight and will drop to 7 parts per million between 2006 and 2011

Corrosion: Is usually measured by immersing a polished copper strip in a fuel for 3 hours at 212° F. This is an indicator of the corrosion expected in copper fuel lines and brass fuel strainers.

Ash Content, Water and Sediment: Ash content is found by burning a given quantity at a very high temperature. The incombustible remains are usually impurities such as rust and sand that are extremely abrasive. Water and sediment are measured together. They are separated from the fuel by a centrifuge. These also cause pump and injector wear.

Safety: Because of the lower volatility, diesel fuels are generally considered to be much safer than gasoline. However, light diesel fuels carry an explosive mixture above the fuel in a closed tank. Except at very low temperatures, the fuel air mixture is too rich to be explosive in a gasoline tank.

Heating Value: Is the actual number of Btu in a sample of fuel. Heating value may be based on weight or volume. Light fuels have a higher heating value by pound, however heavy fuels have a higher heating value by volume. When comparing engine performance on various fuels it is important to evaluate them by heating value of the fuel.

Fuel	Specific gravity	Weight per gallon pounds	Btu/pound heating value higher	lower	Btu /gallon lower
Gasoline	.702	5.86	20,460	19,020	111,457
Kerosene	.825	6.88	19,750	18,510	127,349
Light Diesel	.876	7.30	19,240	18,250	133,225
Medium Diesel	.920	7.67	19,110	18,000	138,060

Diesel Fuel Grades:

No. 1-D Volatile fuel oils from kerosene to intermediate distillates. Used in small high-speed engines with wide variations in load or speed. Also used during very low temperatures.

No. 2-D Fuel oils with lower volatility. Used in high-speed engines under high loads and uniform speeds. Mobile service.

No. 4-D Viscous fuel and blends used in low and medium speed diesels under uniform load and speed.

Grade of Diesel Fuel	Flash Point degrees F	Water & Sediment	Ash Weight %	Viscosity Saybolt , SUS at 100 F		Sulfur Weight %	Copper Strip Corrosion	Cetane Number
	Min	Max	Max	Min	Max	Max*	Max	Min
No. 1-D	100	0.05	0.01	-	34.4	0.5	No.3	40
No. 2-D	125	0.05	0.01	32.6	40.1	0.5	No.3	40
No. 4-D	130	0.5	0.1	45	125	2	NA	30

* Current sulfur values have been dramtically reduced to comply with evolving air pollution standards

54. VEGETABLE AND ANIMAL FAT FUELS:

Vegetable oils have been used in diesel engines for approximately 100 years. Because of their high viscosity, such oils must be heated before use. It is necessary to start, warm up and stop the engine on ordinary diesel fuel. The warm up period is about 5 minutes. Although some engines may operate satisfactorily, there are many problems associated with using straight vegetable oils directly in diesel engines. Typically, problems include:

1. Poor Starting Characteristics.
2. Poor atomization due to the high viscosity of the vegetable oil (approximately 11 to 17 times that of diesel).
3. Coking and formation of carbon trumpets on the injectors.
4. Oil ring sticking
5. Gelling of the lubricating oil due to contamination by vegetable oil.
6. Gum formation due to the drying properties of the oil.

Straight vegetable oils can be mixed with ethanol to improve their properties. Animal fats, due to their high viscosity, can not be used directly in diesel engines. Vegetable oils may be blended with diesel up to 20% oil - 80% diesel.

Conversion of fats and oils to *fatty acid methyl esters* or "biodiesel" reduces the problems associated animal and vegetable fuels. However, increased engine maintenance, including frequent fuel filter changes is specified by all injector manufacturers when using bio diesel blends over 5%. Problems are more prevalent in systems that receive intermittent use such as standby generator sets or seasonal vehicles. Biodiesel blends do not store well and should be used within a few months. All tanks containing biodesel blends should be kept free of water and sediment. Some rubbers, including Nitrile, soften, swell or crack when exposed to biodiesel. Viton is the preferred replacement for affected parts.

Biodiesel blends are specified as Bxx. For example a 20% blend would be B20 and 100% biodiesel would be B100. Typical blends now available are B5, B11, and B 20. B20 and higher blends cloud in cold weather and may be blended with kerosene to reduce clogging of filters.

Biodiesel Concentration % Vol.	Cloud Point Degrees F
0	3
10	5
20	7
30	14
50	18
100	20

ELEMENTARY CHEMISTRY OF BIO DIESEL:

Vegetable oils, animal oils and fats are triglicerides, glycerin bound to 3 fatty acids. If the oil or fat is chemically split by taking up water, the process is called hydrolysis which is Greek for cleavage by water. For each molecule of glycerin that is set free, three molecules of water are taken up, partially to reform the glycerin and partially to reform the fatty acids. This process is often called saponification because it was first observed to take place in the manufacture of soap. Soap is formed by the combination of fatty acid and with a metal such as sodium or potassium. To make soap you would mix a solution of sodium hydroxide (lye) and water and blend it with a transfatty acid or trigliceride (oil or kitchen grease). The glycerin would separate out and the fatty acids would form soap.

Biodiesel is made by a process called transesterification. In transesterification, lye and methanol are mixed to create to sodium methoxide Na+CH3O-. When mixed with waste vegetable oil, the triglyceride is broken into glycerin, methyl esters (biodiesel) and a little soap if there is any water present. The process also works with ethyl alcohol. This process is also called alcoholysis or cleavage by alcohol.

PROPERTIES OF SATURATED AND UNSATURATED FATS AND OILS:

Chemically, fats and oils are made of carbon and hydrogen. If all of the carbon atoms have two hydrogen atoms attached and no double bonds, then it is said to be saturated. This makes the chains of fatty acids straighter and more pliable so that they fit more closely together and harden at low temperatures. However if there are double bonds present and some carbon atoms do not have two hydrogen atoms attached, it is said to be unsaturated and remains liquid at lower temperatures.

Properties of Oils and Esters

Oil Type	Melting Range Deg. F			Iodine Number	Cetane Number
	Raw Oil	Methyl Ester	Ethyl Ester		
Corn	23	14	10	115 -124	53
Cotton Seed	32	23	18	100-115	55
Coconut	75	16	21	8-10	70
Olive	10	21	18	77-94	60
Palm	100	57	50	44-58	65
Rapeseed	41	32	28	97-115	55
Soybean	10	14	10	125-140	53
Sunflower	0	10	7	125-135	52
Lard	97	57	50	60-70	65

When fatty acids are broken from the glycerin they retain their double bonds, therefore the biodiesel retains the properties of the fat or oil it was made from. If the biodiesel is made from lard, it will become cloudy at a higher temperature because the lard is solid at a higher temperature. The more double bonds in the original fat or oil, the lower the cloud point of the finished product.

Many vegetable oils and some animal oils are drying or semi drying and form the base for paints. The drying properties of these oils cause them to form gums or hard films. Drying results from the double bonds in the unsaturated oil molecules being broken by oxygen and forming peroxides. Cross-linking between molecules occurs at this site and the oil permanently polymerizes into a plastic like solid. At high temperatures found in diesel engines, the process is accelerated and the engine can quickly become gummed up.

The drying properties of an oil are described by the Iodine Number. Iodine is added to an unsaturated oil to test an oil for double bonds, or degree of unsaturation. The iodine will attach itself over a carbon atom with a double bond, making a single bond. Therefore, iodine numbers describe the amount of iodine required to saturate or break all of the double bonds in the oil. The amount of iodine in grams absorbed per 100 ml of oil is the iodine number of the oil. The higher the iodine number, the greater the tendency of the oil to form gums and hard films.

Saturated

$$H-C-COOCH(CH_2)_7CH_2CH_2CH_2CH_2CH_2CH_2CH_2CH_2(CH_2)_3CH_3$$
$$|-C-COOCH(CH_2)_7CH_2CH_2CH_2CH_2CH_2CH_2(CH_2)_3CH_3$$
$$H-C-COOCH(CH_2)_7CH_2CH_2CH_2CH_2CH_2CH_2CH_2CH_2(CH_2)_3CH_3$$

Unsaturated

Double Bond →

$$H-C-COOCH(CH_2)_7CH=CHCH=CHCH=CH(CH_2)_3CH_3$$
$$|-C-COOCH(CH_2)_7CH=CHCH=CH(CH_2)_3CH_3$$
$$H-C-COOCH(CH_2)_7CH=CHCH=CHCH=CH(CH_2)_3CH_3$$

314

55. MAKING AN EXPERIMENTAL BATCH OF BIO DIESEL:

The following description is not intended to be an exhaustive description of the waste vegetable oil to biodiesel process but rather an introduction to one type of conversion. While it can be done safely, it is a hazardous process. Lye is highly corrosive and will quickly dissolve your skin or eyes. Methanol is highly flammable. Because laboratory equipment is graduated in SI units, they are used in this description.

1. Heat the waste oil to 100-110F (38 - 43° C) and screen to remove the food particles.
2. Heat the oil to 212° F (100° C) to remove the water. Stirring to prevent water pockets from forming on the bottom or draining them off from the bottom as they form. This is to prevent steam bubbles from splashing hot oil.
3. Mix a solution of 1 gram of lye to 1 liter of distilled water.
4. Mix 10ml of alcohol with 1 ml of de-watered waste oil from above.
5. Add 2 drops of phenolphthalein to the test oil solution.
6. Perform a titration on the oil to determine the amount of lye required to convert it to bio diesel. Using a calibrated burette or dropper, add the lye solution to the test oil, drop by drop. When the solution turns pink for about 10 seconds. Stop adding the solution. Record the amount of solution needed to reach this pH.
7. Calculate the amount of lye needed for the reaction.

lye required = (3.5 grams + titration results) x liters oil

Example: The titration required 2.3 ml to reach the required pH of 8-9. You wish to convert 100 liters of waste oil to bio diesel.

(3.5 + 2.3 grams lye / liter oil) x 100 liters oil = 580 grams lye

Generally, the result will be approximately 6 grams per liter of oil.

8. Using 20% by volume alcohol to waste oil, prepare the sodium methoxide solution by mixing the required amount of lye. The lye and all utensils must be kept dry or soap will form during the transesterification reaction. The first few attempts should be performed on 1-liter batches until you have become familiar with the process. A blender works well for such small batches. Do not use the blender for food after using it for sodium methoxide.

9. Heat the waste oil to approximately 130°F (55°C) and vigorously stir in the sodium methoxide solution. Such vigorous stirring is required for good contact of the reactants and an electric paint stirrer may be used. Stirring should continue for at least 30 minutes. Larger batches may be agitated by a pump. See drawing of settling tank in the oil filtration chapter.

10. Allow the mixture to settle for two hours while keeping it at 100°F. When the mixture has separated, drain off the glycerin into one container and the bio diesel into another.

11. Add a little vinegar to the bio diesel wash water to neutralize the lye. This is usually at a rate of 1 ounce per 25 gallons or 30 ml per 100 liters. Add the biodiesel to an equal amount of water and vigorously stir it for 5 minutes or use the paint stirrer. Allow it to settle overnight and decant. Repeat the process 2 more times. Vinegar is not needed after the first wash. The finished oil should have a pH of 7. The oil may be heated to drive off any residual water. The bio diesel should be filtered through a transparent or glass type fuel strainer and a standard diesel fuel filter before use.

Larger containers, immersion heaters and pumps may be employed for larger operations, greatly reducing the amount of labor involved.

If you choose to prepare your own biodiesel, work on small batches until you have mastered the process. Most of the failures are related to poor control of the amount of lye used. Pay close attention the titration.

56. OIL FILTRATION AND CLARIFICATION:

Gasoline, lubricating and fuel oil are filtered to remove dirt, water, metal particles and other foreign matter. By removing these foreign substances, acid and sludge, which result from oil being mixed with this foreign matter, are also reduced.

Darkening of oil is mainly due to the introduction of fine carbon, which stays in "colloidal suspension." Centrifugal clarification removes all impurities except colloidal carbon.

TYPES OF CLARIFICATION: Oil cleaning is typically achieved by two methods, precipitation and or filtration. Precipitation includes: Gravity –Settling, Forced- Centrifugal and Chemical precipitation.

Filtration includes Strainers, Pressure filtration and Coagulation.

Usually combinations of several of the above methods are used. Strainers are used for preliminary cleaning and better systems employ parallel strainers so that one may be cleaned while the other is in use.

Settling by gravity is a slow process and will not remove fine carbon particles or sludge. The only advantage to settling is its cheapness. Typically the oil is pumped into a large tank and left undisturbed for at 10 days to 2 weeks. Even a slight addition of oil to the tank destroys the necessary quiescence. When this system is employed, it is best to have 2 tanks and alternate the discharge from one to the other as the companion tank is resting. Impurities will not readily settle out of fuels or oils with viscosity of 150 SSU or higher at the storage temperature. Such oil is heated by a hot water coil to 180 or 200° F, provided this is safely below the flash point.

When oils contain more than about .1 percent sediment, it is advisable to purify them by a centrifuge. Two centrifuges in series may be used, one as a purifier to remove water and the second as a clarifier. When cleaning heavy oils by centrifuge, the oil is heated to temperatures above 150° F. A hot water coil is located around the suction pipe in the storage tank.

The hot water pipe should have no joints within the tank. A hot water jacket may also heat the fuel line. Since the fuel may be heated nearly to the gassing point, centrifuges for heated fuel oil should be capable of being closed vapor tight.

CENTRIFUGAL PRECIPITATION:

When a mixture of oil, dirt and water stands undisturbed, gravity tends to separate the liquid into an upper layer of oil, an intermediate layer of water and a lower layer of dirt or solid material. When the mixture is placed in a rapidly revolving bowl, centrifugal force accelerates the separation. Solids collect upon the bowl, water forms an intermediate layer and clean oil, being the lightest constituent, moves to the center of the bowl. The discharge holes of the bowl may be arranged so that water can be drawn off and discharged. The solids are cleaned from the walls as required.

The speed of rotation, viscosity of the oil and the speed of the oil flow through the centrifuge affect the quality of separation. Practical experience dictates the speed of oil flow through the centrifuge for proper cleaning. More material is removed by lower flow rates. If the feed is too

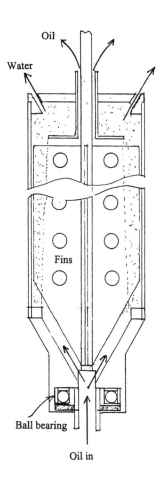

318

rapid, only partial clarification will result. Heating the oil to reduce the viscosity is one of the most effective methods of improving purification.

A centrifuge may be operated as a purifier or a clarifier. If the oil contains water, then a purifier is used. The purifier has two sets of discharge holes. Water is discharged from one set and oil is discharged from another. If the oil contains only solid matter, then the centrifuge is set up as a clarifier by covering the water discharge holes leaving with only the oil discharge holes open.

Inside the bowl, a partition of radially spaced flat plates rotates with the bowl forcing the oil to spin at the bowl speed. A cone at the bottom of the bowl brings the speed of the liquid up smoothly so as not to form an emulsion.

Typically, oils are cleaned at 2000 to 2500 G's with 1 G being equal to the force of gravity. The G force is dictated by both the bowl diameter and RPM . G force, bowl speed and diameter may be determined by the following:

$$f = \text{d RPM}^2 / 265^2 \qquad \textbf{RPM} = 265\sqrt{f\text{d}} \,/\, \text{d} \qquad \textbf{d} = 265^2 \, f \,/\, \text{RPM}^2$$

Where f = force in G's and $\textbf{d}$ = diameter in inches

Example: find the RPM for 2250 G's using a 3-inch diameter bowl.

$$\textbf{RPM} = 265\sqrt{f\text{d}} \,/\, \text{d}$$

$$\text{RPM} = 265\sqrt{2250 \times 3} \,/\, 3$$

RPM = 7257

7257 RPM is nearly 2 times 3600 RPM, the standard speed of 2 pole electric motors in the US. It should be an easy matter to size a set of pulleys for this speed.

CALCULATING PULLEY SIZES:

Obtaining the correct centrifuge speed from the motor speed is calculated from the ratio:

$$RPM_1 / P_2 = RPM_2 / P_1$$

RPM_1 = Motor speed P_1 = Motor Pulley Diameter

RPM_2 = Fan speed P_2 = Centrifuge Pulley Diameter

The formula may be easily manipulated to solve for either pulley size if one pulley size and the speeds of the motor and centrifuge are known.

$$P_1 = (P_2 \times RPM_2) / RPM_1$$

$$P_2 = (P_1 \times RPM_1) / RPM_2$$

Calculate the pulley diameter for a 3600-RPM motor if the desired centrifuge speed is 7257 RPM and the centrifuge pulley is 2 inches in diameter:

$$P_1 = (P_2 \times RPM_2) / RPM_1$$

$$P_1 = (2 \times 7257) / 3600$$

$$P_1 = 4.032 \text{ inches diameter}$$

OIL CLARIFICATION BY CHEMICAL TREATMENT:

Chemical treatment is the most expensive of the clarification methods and is used for precipitation. Extremely fine particles of carbon or other fine matter are removed by uniting them with a coagulating agent. Brief but violent mixing of hot oil and a solution of hot water containing soda ash causes the carbonaceous matter to be precipitated in the form of sludge between the oil and water. The oil and water are heated to at least 180° F before mixing. After sufficient mixing, the treated oil is allowed to stand overnight.

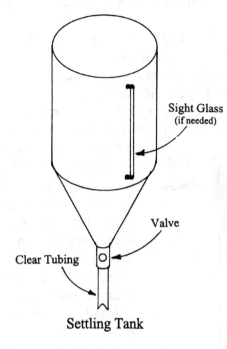

Sight Glass (if needed)

Valve

Clear Tubing

Settling Tank

STRAINERS AND PRESSURE FILTERS:

Because condensation readily forms in gasoline and oil tanks, filters for all liquid fuels should include an oil-water separator. A fuel strainer should be used before an inline filter.

Filters are typically rated by type of fluid filtered, such as fuel or hydraulic oil. Filters are also rated by the flow rate in gallons per minute (GPM) and micron. Micron refers to the size of particle that the filter traps. A Filter marked 10 microns has some ability to capture particles that are as small as 10 microns however you will not know exactly what this means unless you have some information about how the filter was tested.

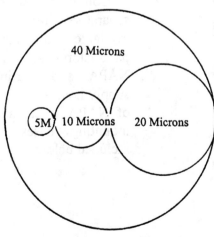

40 Microns

5M 10 Microns   20 Microns

40 Microns = .00156 inch

Nominal Micron Rating means that a filter can capture a percentage of the particles of a certain size for example a filter may be rated at 90% 10 micron.

The Absolute Micron rating is a single pass test. Particles that pass through the filter are measured and counted.

Multi Pass Beta Rating is used for power applications such as transmissions, power steering and hydraulic machinery.

Filters should be selected not only by their micron size but also by the proper application. Newer common rail diesel injection systems are more susceptible to wear and clogging from fine particles than the older type systems. Older systems may require a 10-micron filter, however new systems may require as small as a 2 micron filter.

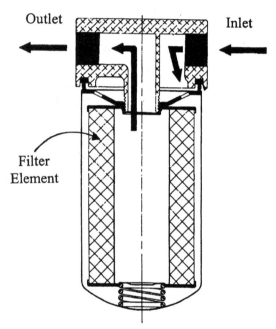

Outlet

Inlet

Filter Element

Spin on Filter

Commercial water separators-filters are available from auto parts stores such as NAPA and marine supply houses. A list of NAPA filters and mounting bases is included below.

NAPA Fuel Filters and Mounting Bases

Part Number	Description	Dimensions	Micron Rating	Filter Thread	Threaded connections
4762	Filter Base	L=4, W=3.125, H=2.5	NA	3/4-16	1/2 NPT
4764	Filter Base	L=4,W=3, H=2.5	NA	3/4-16	1/2 NPT
4768	Filter Base	L=4. W=3, H=2 7/8	NA	1-12	1/2 NPT
4770	Filter Base	L=4. W=3, H=2 7/8	NA	1-14	1/2 NPT
4309	Filter Base	L=3 1/2 H=3 1/4 W=3 3/4	NA	1-12	1/2 NPT
4001	In Line Filter Base	L=3.74 OD = 2.25	NA	1-12	1-NPTF
4034	In Line Filter Base	L=3.74 OD = 2.25	NA	1-12	3/4-NPTF
3382	Fuel Filter	H=3.812, OD=3.675	10	3/4-16	
3386	Fuel Filter	H=3.197, OD=3.25	10	3/4-16	
3393	Fuel Filter	H=4.069 OD=3.25	10	3/4-16	
3405	Fuel Filter - water separator	H=7.22 OD=3.7	12	1-14	
3418	Fuel Filter - water separator	H=8.42 OD=3.7	12	1-12	
3504	Fuel Filter	H=7 OD=3.66	10	1-14	
3522	Fuel Filter with Drain	H= 7.45 OD = 3.78	10	1-14	
4006	Fuel Filter	H=5.2 OD=3.7	10	1-12	Max Flow 15GPM

57. BATTERIES:

A battery transforms chemical energy into electrical energy at the expense of the electrode, which goes into solution.

If two lead strips are immersed in dilute sulfuric acid and a DC current is put between them, after a short time, one plate will turn dark and the other will appear unchanged; however upon close inspection, its surface will be spongy. While the current is supplied to the cell, the voltage across the plates is about 2.5 volts. If the current is turned off, the voltage across the plates is about 2.05 volts. The cell is now able to supply a small current. As energy is delivered by the cell, the voltage drops off and the dark brown plate changes back to nearly its original color.

The dark plate is the positive plate and the spongy plate is the negative. Applying current to a cell converts the positive plate to lead peroxide however the negative plate is not changed, except to a spongy form of lead. When the cell delivers energy, the lead peroxide is converted to lead sulfate and the spongy lead of the negative plate is also converted to lead sulfate so that they become chemically equivalent. If two plates are chemically equivalent, then no voltage exists between them.

When a cell is being charged, the voltage is about 2.5 volts, however the cell is only able to deliver about 2.05 volts. The .45 volt difference is due to the internal resistance of the cell. Every battery has internal resistance, which reduces the flow of current and the terminal voltage. The resistance is in the electrodes, the electrolyte and the contact surfaces between the electrodes and electrolyte. Therefore, larger electrodes have less resistance and can carry more current. The resistance in a cell may also be reduced by spacing the electrodes closer together. This reduces the current path inside the cell. Increasing the size of the cell increases the current capacity but not the voltage. The voltage depends only upon the electrode material and the electrolyte. Therefore two lead acid batteries, one very large and one very small, would have the same terminal voltage.

CHEMICAL REACTIONS IN LEAD ACID BATTERIES: The chemical reactions that take place in a storage cell are:

Battery Discharged				Battery Charged		
(+ Plate)	(- Plate)			(+Plate)	(-Plate)	
$PbSO_4$ +	$PbSO_4$ +	$2H_2O$	$\leftrightarrow$	PbO_2 +	Pb +	$2H_2SO_4$
Lead sulfate +	Lead Sulfate +	water	is changed to	Lead peroxide +	Lead +	Sulfuric acid

When the battery is being charged, the only changes that takes place in the electrolyte is that water is converted into sulfuric acid. This accounts for the increase of specific gravity of the electrolyte. Upon discharge, the sulfuric acid is dissociated and reacts with the lead peroxide to form water. During discharge, the specific gravity of the electrolyte decreases. Free hydrogen is given off by the negative plate and oxygen at the positive plate during charging. Hydrogen is explosive and must be ventilated.

BATTERY CONSTRUCTION:

Because of the flaky nature of lead peroxide and due to the small surface area, it is not practical to make batteries from lead sheets. Some batteries have grooved plates where the lead forms fins while others have rolls of corrugated lead ribbon. These plates are called Plante plates. Such plates are usually designed for 1800 to 2500 charge and discharge cycles before their capacity fall to 80%.

Plates may also be of a lattice type with a lead oxide paste being applied. The battery is then charged. The paste on the positive plate is converted to peroxide and the negative plate is converted to spongy lead. Pasted plates are used when a light compact battery with a high short-term energy capacity is required, such as engine starting. Fiberglass mats are applied to the plates to reduce their erosion, extending their life. Pasted plates have a life of about 750 charge-discharge cycles before they are reduced to 80% of the initial capacity.

Six and twelve-volt batteries are made by connecting several cells producing approximately 2.05 – 2.1 volts in series.

ELECTROLYTE:

The electrolyte in a battery must be chemically pure, therefore only distilled water is used is used to dilute the sulfuric acid. The concentration of the acid is determined by the specific gravity of the solution. From a standpoint of life, a lower specific gravity is used because there is less action on the surfaces of the plates.

A high concentration or specific gravity increases the output of the cell. However, when a cell is fully charged, the specific gravity should never exceed 1.300. To prevent a rapid drop of current during discharge, the specific gravity at the end of the discharge

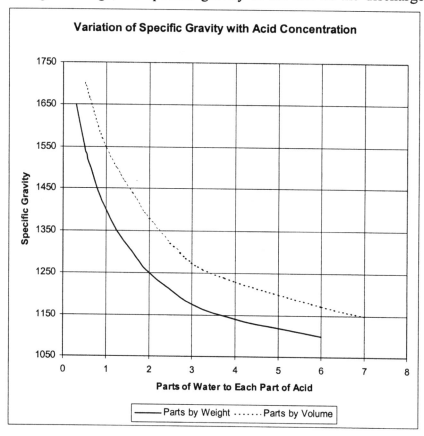

Variation of Specific Gravity with Acid Concentration

cycle should not be less than 1.100. Stationary batteries operate with the electrolyte range between 1.210 and 1.150 while small portable batteries range of operation is between 1.280 and 1.125.

The specific gravity of a solution is determined by a hydrometer consisting of a weighted bulb with a graduated tube. Depending upon the concentration of the acid, the bulb floats at different heights and the specific gravity is read at the point where the acid intercepts the tube.

CHANGES IN SPECIFIC GRAVITY DURING CHARGING:

When a battery is charged, oxygen is given to the positive plate to convert it to lead peroxide and sulfate ions SO_4 are given off at the negative plate to leaving spongy lead. The oxygen is supplied by the water leaving the hydrogen, H_2, to combine with the SO_4 forming sulfuric acid, H_2SO_4. The specific gravity of the solution is a good indicator of the charge of the cell. If the hydrogen and oxygen escape from the cell during charging, then the equivalent water is lost.

Open Circuit Battery Voltage			
% Charge	Specific Gravity	6 Volt	12 Volts
100	1.277	6.37	12.73
90	1.258	6.31	12.62
80	1.238	3.25	12.50
70	1.217	6.19	12.37
60	1.195	6.12	12.24
50	1.172	6.05	12.10
40	1.148	5.98	11.96
30	1.124	5.91	11.81
20	1.098	5.83	11.66
10	1.073	5.75	11.51

State of Charge Relative to Specific Gravity and Open Circuit Voltage

BATTERY RATINGS:

Batteries are rated based on an 8-hour discharge. However, many are specified at 5 and 20 hours discharge rates. Therefore, if a Plante battery can deliver 30 amps continuously for 8 hours it has a rating of 30 x 8 = 240 amp-hours. The normal charging rate of the battery would also be 30 amps. A battery can deliver a higher rate of discharge for a short time, however the discharge rate will be some fraction of the 8-hour rate. This decrease in capacity is due to the in ability of the free electrolyte to rapidly penetrate the pores of the plates. Higher temperatures increase the electrolyte's ability to flow through the pores in the plates. After a deep discharge, a battery that has stood for a short time allows the plates will recover somewhat and the battery can again deliver current. Batteries are able to discharge at extreme rates for a short amount of time, which is why a starting battery rated at 12.5 amps for 8 hours is able to deliver 450 cranking amps.

Percentage Capacity with Various Discharge Rates:						
Discharge Rate	Hours				Minutes	
% Capacity at 8-hr rate	8	5	3	1	20	6
Plante Plates	100	88	75	55.8	37	19.5
Pasted Plates	100	93	83	63	41	25.5

COMMERCIAL BATTERIES FOR POWER SYSTEMS:

Batteries include starting or automotive type and deep cycle types. Automotive (starting) batteries typically see about a 3% discharge before they are recharged by the engine's alternator. Deep cycle types are designed for a 50 to 70% discharge before recharging. A starting battery will quickly be ruined if used in a deep cycle application. Battery life is affected by the depth of discharge with discharge cycles over 50% greatly reducing the battery life.

Inverter power systems output 120 or 240 volts and are powered by deep cycle storage batteries. Such batteries may be stationary or mobile. Deep cycle stationary batteries are assembled from 2-volt cells and provide very long running cycles. These cells are typically rated from 1000 to 2000+ amp-hours and last 10 to 20 years. Mobile batteries include automotive and RV types, forklift type, golf cart and marine type. Generally, 6-volt deep cycle "renewable energy" or deep cycle golf cart batteries are used for small power systems. They are readily available and not too expensive. Automotive and RV batteries (pasted plates) are not deep cycle and are not suitable for power plants. Inexpensive marine "deep cycle and starting" batteries are often automotive type with combination terminals. The smallest deep cycle battery for inverter application should be at least 250 amp-hours.

BATTERY FAILURE:

A battery ages as the material on the positive plate flakes off due to the normal expansion and contraction that occurs during cycling. This causes a gradual loss of plate material and a build up of the flakes in the bottom of the battery where it forms sludge. If the sludge builds up high enough to reach the plates, they short out, discharging the battery. In similar situation, as the positive plate decays, eventually there is not enough material to support the discharge current.

Approximately 85% of batteries fail due to sulfation. During the normal cycling process, soft lead sulfate crystals are formed on the plates. When a battery is left in a discharged condition, is continually undercharged or the electrolyte levels are below the tops of the plates, the soft lead sulfate forms permanent hard crystals than can not be reconverted during charging. The longer the battery is let in this condition the larger and harder the crystals become. Eventually there is not enough free plate material to support the electrical load. Keeping a battery fully charged prevents hard lead sulfate from forming.

Approximately 50% of battery failures are caused by water loss due to neglect or overcharging. Plates that are exposed to the air corrode and form hard lead sulfate crystals. In a discharged condition, the water should just cover the plates. In the charged condition, the water level should be 1/8 inch below the filling well.

BATTERY CHARGING:

The maximum charging rate in amperes may be equal the battery's ampere-hour capacity divided by 8. For example, the charger for a 225 ampere-hour battery should be 225 / 8 = 28 amps. For battery banks, use the ampere-hour rating of the entire bank. Smaller chargers may be used however the charging time will be increased.

Many types of chargers are available. They are usually rated by their beginning charging rate or the maximum output available at the beginning of the charge cycle. Automatically regulated or programmable chargers are preferred.

Voltage Setting	System Voltage				
	6V	12V	24V	36V	48V
Daily Charge	7.2 - 7.4	14.4 - 14.8	28.8 - 29.6	43.2 - 44.4	57.6 - 59.2
Float	6.6	13.2	26.4	39.6	52.8
Equalize	7.8	15.5	31.0	46.5	62

EQUALIZING:

Equalizing is an overcharge performed after a battery is fully charged. It breaks down sulfate crystals and mixes the electrolyte when the acid concentration is greater at the top of the cells. Equalization is performed from one a month to twice a year, or when the specific gravities vary ± .015 on a fully charged battery.

To equalize a battery, check and record the specific gravities of the cells. Set the charger for the equalizing voltage and begin charging. The battery will begin bubbling and gassing. Check the specific gravities every hour until they no longer increase. Equalization is then complete.

BATTERY SAFETY:

Because batteries contain acid and have the potential to explode, goggles and gloves should always be worn when performing any service on a battery.

Hydrogen gas is generated as the cell is charged and comes from the negative plate. Oxygen is also generated and comes from the positive plate creating a very explosive mixture. Any spark can cause a battery explosion and subsequent spattering of acid, potentially blinding anyone close by. Never plug in a battery charger and then connect it to the battery. Likewise, never jiggle the cable connections when the battery is charging and always unplug the charger before removing the cables. Never allow a spark or flame near a battery.

Over charging dissociates the water in the cell into hydrogen and oxygen gas. It also causes the plate material to flake off.

Do not install a bank of mixed batteries of different capacities or ages. Replace all the batteries at once.

Always remove any loose jewelry that might make contact with an exposed terminal.

GRAVITY TESTING:

Bulb type hydrometers are inexpensive and available at your battery supplier or at auto supply stores. To check the specific gravity: Do not add water prior to testing. Fill and drain the hydrometer about 3 times before drawing a test sample. Draw enough electrolyte to fully support the float. Record the reading and return the electrolyte to the cell. Replace the caps and wipe any spilled electrolyte with a cloth dampened with baking soda and water. Do not get any baking soda in the cell. Correct the readings to 80° F by adding .004 for every 10° F above 80° F or subtracting .004 for every 10° F below 80° F.

US Battery Model	Amp Hours (20hr. rate)	Minutes @ 75amps	Minutes @ 25amps	Length	Width	Height	Weight	Pallet Qty
US-250	250	150	540	11-5/8" (295)	7-1/8" (181)	11-5/8" (295)	72 lbs.	36
US-250HC	275	168	600	11-5/8" (295)	7-1/8" (181)	11-5/8" (295)	78 lbs.	36
US-250E	222	132	477	11-5/8" (295)	7-1/8" (181)	11-5/8" (295)	64 lbs.	36
US-305	305	175	675	11-7/8" (302)	7-1/8" (181)	14-5/8" (371)	87 lbs.	18
US-305HC	335	203	747	11-7/8" (302)	7-1/8" (181)	14-5/8" (371)	96 lbs.	18
US-305E	286	172	622	11-7/8" (302)	7-1/8" (181)	14-5/8" (371)	85 lbs.	18
L-16	375	214	810	11-7/8" (302)	7-1/8" (181)	16-3/4" (425)	111 lbs.	18
L-16HC	415	236	890	11-7/8" (302)	7-1/8" (181)	16-3/4" (425)	117 lbs.	18
L-16E	353	187	753	11-7/8" (302)	7-1/8" (181)	16-3/4" (425)	104 lbs	18

US -Battery Co. 653 Industrial Park Dr. Evans, GA 30809   (706) 868-0533   www.USBattery.com

58. INVERTERS:

Generators are most efficient when operating at their rated capacity, however household loads vary widely and most often are much less than the rated capacity of the generator. Keeping a generator fully loaded by charging batteries and then running off of the batteries has been done at least since the earliest submarines and is the same technology used in many of the new diesel electric cars. If all of your appliances operated on direct current, then you would not need much more than the batteries and a regulated charger.

An inverter converts direct current from a battery to alternating current for use in home appliances. All inverters are not created equal. There are true sine wave inverters, modified sine wave and square wave inverters. True sine wave inverters produce the same type of power as you get from your wall socket. Therefore, a true sine wave inverter, provided it is large enough, will power any appliance as if it were plugged directly into utility power.

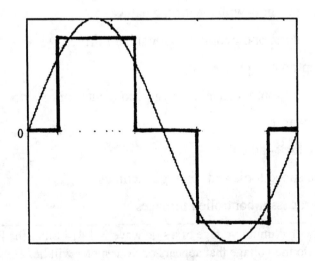

Modified Sine Wave and Sine Wave

Modified sine wave inverters are less expensive and more common. While they will power most appliances, such as freezers, microwaves, and toasters, a few electronic devices will not operate properly on a modified sine wave. Some electronic devices such as variable speed tools, battery chargers, cell phones, lighting dimmers and certain electronic power supplies depend upon switching on during a certain point in the sine wave. In the modified sine wave, the voltage is either at the maximum or at zero with no values in between. Switching for voltage control is done by an electronic switch called an SCR or Triac. A circuit detects when the voltage passes through zero and then turns the SCR on at the proper time for the required voltage. This all happens very quickly as one cycle only requires 1/60 of a second or .0167 second. Such switching circuits work well with the true sine wave output, however some may not work with the modified sine wave. Equipment that may not operate properly with a modified sine wave output:

Fluorescent lights with electronic ballast

Laser printers, photocopiers, optical hard drives

Some laptop computers

Medical Equipment such as oxygen concentrators

Battery chargers

Digital Clock-radios

Variable speed tools and sewing machines

Micro-processor controlled furnaces

Note: The maximum voltage of a sine wave is 1.414 times the RMS voltage. So the voltage that appears on your meter will be: 169 / 1.414 = 119.5 volts or about 120 volts. Because the sine wave's voltage rises and falls with time, it requires 169 peak volts to equal the heating value of 120 DC volts.

You can see how sine wave power is able to supply variable voltage by the following example:

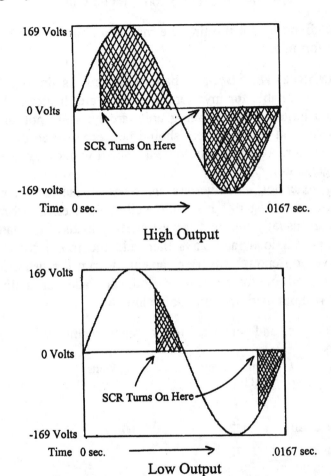

High Output

Low Output

Voltage (V) at any point in time for 60 Hz power is:

$$V = V_{max} \text{Sin}(21600\ t) \quad t = \text{time in seconds}$$

Example: Find the voltage at .0073 seconds

$$V = 169 \sin(21600 \times .0073) = 64 \text{ volts}$$

Current is found by substituting the maximum current for V_{max} in the above formula.

SIZING AN INVERTER: Because battery capacity is limited, some loads are not suitable for inverters. Electric water heaters, electric ranges, and large air conditioning units are typically not suitable loads. Stoves and water heaters should be gas powered or driven by the generator. AC units should be the small window type or run from the generator.

Like a generator set, inverters are sized by the largest motor load to be started. Usually this is the well pump. Refrigerators and AC units are usually the next largest starting loads. Calculate the typical expected load, add 25% then add the motor load. This should give you enough inverter capacity to run the usual loads, start your well or freezer and periodically add an additional appliance without having to turn something off.

Example: Size and inverter for a small cabin with a few appliances:

	Running Watts	Starting Watts
Microwave:	1100	
TV	300	
Refrigerator (small)	132	400
Lap Top Computer	75	
Lamps	200	
Small AC	750	2250
	2557	
Times 1.25 =	3196 watts	
Well pump ½ hp	1176*	3528

*The well pump is approximately the same load as the microwave. If both are not running at the same time along with everything else, then this capacity does not need to be duplicated. Load management is important for small systems.

The approximate total watts = 3200

$$3200 \text{ watts} / 120 \text{ volts} = 26.67 \text{ amps}$$

This is between 25 and 30 amps, going to the nearest larger size, a 30-amp 3600 watt inverter should work. Assuming that we will need about 6,000 surge watts to start the pump and run the typical load, check the surge watts of two inverters:

Outback 3600-watt sine wave inverter, surge watts   = 6,000

Xantrex DR 3600-watt modified sine wave, surge watts = 7,500

Either inverter will work. I will have to check my lap top and cell charger to be sure that they will operate on a modified sine wave. Checking a few dealers, there is a few hundred dollars difference between the two inverters.

SIZING A BATTERY BANK: Assuming a 24-hour cycle, the required amp-hours is determined by multiplying the run time of the appliance by the running watts. The total watt-hours is then divided by the battery bank voltage to find the required amp-hours.

	Running Watts	Running Time, hours	Watt-hours
Microwave:	1100	.5	550
TV	300	8.	2400
Refrigerator (small)	132*	8*	1056
Lap Top Computer	75	2	150
Lamps	200	8	1600
Small AC	750	10	7500
Well pump ½ hp	1176	.75	882
		Total :	14,138

*The refrigerator runs only about 1/3 of the time or 8 hours per 24 hour day

337

The total watt-hours are:14,138. Divide by the battery bank voltage to get amps. Choosing a 48 volt bank for the Outback inverter:

14,138 / 48 volts = 294 or approximately 300 amp-hours

To ensure long battery life, we do not want more than a 50% discharge, double the amp-hour rating to 600 amp hours.

The battery bank should be 48 volts and at least 600 amp hours.

Checking battery specifications, 16   #305HC will do the job. However #L16 batteries provide a little extra capacity for battery aging, inverter, battery and cable losses.

That is a pretty good size battery bank. I can reduce the size of the battery bank, and possibly the size of the inverter by running the generator in the evenings to power the cabin and charge the batteries. Another good investment might be a window AC with a thermostat. The one I am currently using has an on/off switch. That single change may reduce my battery bank capacity by 160 to 200 amp hours. The cost savings in batteries will pay for the new unit. I can also cycle the generator twice per day. Once in the morning when it gets hot enough to run the AC and again in the evening when the heavier electrical loads are on. While the utility grid seems to be an endless supply of power, timing and management of electrical loads is the key to successful operation of an inverter system.

RUNNING 240 VOLT SYSTEMS:

Typically, houses are wired for 240/120 volts. In order to reduce the wire size and electrical losses, well pumps and larger appliances are usually wired for 240 volts. If the amount of 240-volt equipment is limited, you may add a transformer in line to step up the voltage to the appliance. However, if many circuits are wired for 240 volts, multiple inverters can be wired in series for 240 / 120 volt operation. Both Outback and Xantrex offer "stackable" inverters. Outback Inverter Systems allows parallel-series wiring of inverters for a total of 36,000 watts. This is a great advantage in that it allows you to start with one or two inverters and expand your system later.

BATTERY WIRING:

Voltage drop is relative to the resistance of the wire and the current. The low voltage side of the inverter operates on high currents. Therefore large cables are required to prevent an unacceptable voltage drop. Cable length for 12-volt systems should not exceed 5 feet, 24 and 48-volt systems should be kept to 10 feet or less. Longer runs require parallel cables (multiple positive and negative cables). Higher input voltage allows more power to be drawn with less current and less voltage drop. A 1-volt drop in a 12-volt system equals an 8.33% drop, while a 1-volt drop in a 48-volt system is only 2%. Voltage drop for cable length is calculated as seen in the feeder cable chapter.

Battery Cable - Maximum Capacity		
Cable size AWG	Capacity - amps, in conduit	Capacity - amps open - no conduit
# 2	115	170
# 2 / 0	175	265
# 4 / 0	250	360

The ignition coil on an engine depends upon a changing magnetic field to induce high voltage in a coil of wire. The changing flow of current in the inverter's battery cables sets up a changing magnetic field that induces voltage spikes in the battery cables. These spikes must be absorbed by the filter capacitors in the inverter. The magnetic field also opposes the flow of current through the battery cables. The force is relative to the size of the coil or loop, with larger loops generating higher forces. The two battery cables in an inverter system form a single loop. If these cables are taped together, the magnetic fields tend to cancel each other, greatly reducing the inductance or opposing forces and the induced voltages. Therefore, battery cables should be kept as close together and as short as possible.

OTHER CAUSES OF VOLTAGE DROP:

Although a battery cable has a poor connection or many broken strands, when the resistance is checked with an ohmmeter, it may read 0.0 ohms. Yet, when a heavy current flows through the cable, the resistance of the cable causes a large voltage drop. If the inverter requires a surge of 200 DC amps to start a small motor and the resistance of the cable is .01ohm, which is too small to be detected by most ohmmeters, the voltage drop will be:

$$200 \text{ x } .01 = 2 \text{ volts}$$

In a 12-volt system, this is about a 16% drop. If the resistance of the connection is .02 ohm, the voltage drop would be 4 volts.

The best way to check for such small resistance is to check the voltage drop across the cables with a digital voltmeter when the system is fully loaded.

I use welding cable for my low voltage applications. The cable is very flexible due to the very fine strands of wire, however the strands are very delicate. In order to prevent breaking the strands, I wrap the ends with copper foil, which is available from welding supply houses, before the cable ends are clamped.

Cable resistance increases with temperature. The increase in resistance may be approximated by:

$$R_t = R_1 (1 + .00236 \text{ t})$$

R_t = resistance at new temperature, R_1= resistance at 77°F

t = new temperature – 77°F

Example: Find the increase in resistance for 100 feet of #10 wire if the temperature is 150°F, seeing page 33, 100 feet of #10 =.102 Ω, where Ω = ohms, and 150 –77 = 73

$$R_{150} = .102 (1 + .0023 \times 73) = .119 \, \Omega$$

Although the increase appears to be small, larger currents would see a considerable voltage drop. The same situation exists for battery cables.

BATTERY ROOM VENTILATION:

Batteries generate hydrogen sulfide, hydrogen and oxygen gas during the cycle. Hydrogen sulfide is corrosive and hydrogen is explosive. These gasses must be ventilated. In a small battery enclosure, passive ventilation such as a 1.5-inch hole in the bottom of the enclosure and a hole in the top of the enclosure allows the hydrogen and warmer air to rise and escape the battery compartment. Holes should be screened to prevent the entry of rodents. A Small AC or DC brushless cooling fan may be used for forced ventilation of large banks. The fan should be installed at the bottom of the enclosure and pressurize the battery compartment so that it will not be drawing in a corrosive or explosive mixture. These little square fans produce 1.5+ cfm and draw as little as .5 watt. A brush type motor should not be used due to the sparking nature of the brushes.

CORROSION PROTECTION:

The exposed cable ends should be protected from spilled or leaking acid attack by use of the appropriate felt terminal pads or a mixture of baking soda and petroleum jelly or grease. The enclosure bottom should get a light layer of baking soda to neutralize any leaking acid.

SERIES AND PARALLEL WIRING OF BATTERIES:

When wiring batteries in series, voltages add. Current remains the same. Assuming each battery below is 6 volts and 100 amps:

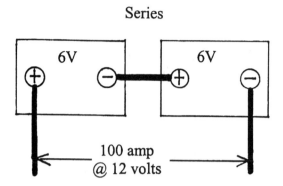

When wiring batteries in parallel, current capacity adds, voltage remains the same:

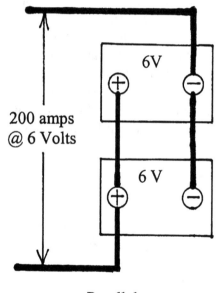

Parallel

Batteries may also be wired in series "banks" to raise the voltage and then two or more banks may be wired in parallel to increase the amperage.

CONCLUSION:

While I thought that I might be able to build a home generator that would work, I was surprised at how well it works. The completed project exhibits very good frequency control, is very quiet and generates clean power that operates everything in my house, including my 4-ton AC unit as well as if I were connected to utility power.

The propane generator exceeds most of my expectations. The one area where it fell a little short was in the total output, however due to the small carburetor and lower volumetric efficiency, that might have been expected. Using two electric ranges connected in parallel and a clamp-on ammeter the output was as follows: 50 amps at 240 volts-60 Hz, 55 amps at 240 volts-57 Hz, 60 amps at 220 volts-55 Hz. Checking published data from various generator builders, the output per cubic inch of engine displacement is right in line with industry results. Therefore, I am not disappointed. Extrapolating from these results, I would assume that you can expect 125 to 135 watts per cubic inch of engine displacement at 1800 rpm. Therefore, a 2-liter engine would be good for 15,250 to 16,500 watts.

Gasoline to propane conversion is not difficult and works so well that I have considered performing a conversion on one the Nissan Sentras in my driveway.

The remote starter is an impressive little device that allows me to start and stop the genset from inside my house. I am very pleased with its operation.

The engine silencing techniques work well and the addition of an input muffler to the diesel engine dramatically reduces the noise. Removing the engine fan and going to shell and tube heat exchangers further reduced the diesel noise to a low hum, even before the addition of an enclosure.

Although long term tests of the heat recovery system, corrosion protection system and various mixtures of waste oil have not been performed, short run results look good. Mixtures of 30% waste oil mixed with salvaged heating oil starts and runs the diesel set with no trouble. In the future, a fuel oil heater will be built and tested, hopefully allowing me to use an even lower quality fuel successfully.

Summing it all up, I am extremely pleased with the results.

Looking forward, the idea of ammonia absorption cooling is intriguing. Such coolers were common in households during the 1930's. There is plenty of exhaust heat available and such a device would certainly complement the project.

Much information is contained in this book and it should allow you to address any situation in your small power generation project.

Best of luck to you,

Stephen D. Chastain

Bibliography

American Gas Association, *Combustion: A Reference Book on Theory and Practice*, 3rd ed. (New York: American Gas Association, 1932).

Black, Newton Henry, *An Introductory Course in College Physics*, (New York: The Macmillan Company, 1950).

Bleier, Frank P., *Fan Handbook*, (New York: McGraw-Hill Book Company, Inc., 1998).

Brown. William H., *Organic Chemistry*. (New York: Harcourt Brace 1995).

Cain, Tubal, *Spring Design and Manufacture*, (Great Britain: Argus Books, 1988).

Cengel, Yunus, A., and Michael A. Boles, *Thermodynamics: An Engineering Approach*, 3rd ed. (McGraw-Hill Companies, Inc., 1998).

Dawes, Chester L., *A Course in Electrical Engineering Vol. 1 & 2* (New York: McGraw-Hill, 1934 & 1947)

Detroit Diesel, *Bulletin 50. Cooling Guidelines for Radiator Cooled Engine Applications*. (Detroit Diesel, 1993).

Diesel Engine Manufacturer's Association, *Standard Practices for Low and Medium Speed Stationary Diesel Engines*. (Chicago, 1951).

Fontana, Mars G, *Corrosion Engineering*, (New York, McGraw-Hill, 1967).

Fraas, Arthur P. *Heat Exchanger Design*, (York: John Wiley & Sons, Inc., 1989).

Fox, Robert W., and Alan T. McDonald, *Introduction to Fluid Mechanics*, 5th ed. (New York: John Wiley & Sons, Inc., 1998).

Frylunn, Marvin M. *Lightening Protection for People and Property* (New York: Van Nostrand, 1993)

Geyer, Wayne, *Handbook of Storage Tank Systems*, (New York: Marcel Deker. Inc, 2000).

Goldberg, David E., *Fundamentals of Chemistry*, (Wm. C. Brown Communications, Inc., 1994).

Harris, Cyril M., Ph.D., *Handbook of Noise Control*, (New York: McGraw-Hill Book Company, 1957)

Hay, Nelson E., *Guide to Natural Gas Cogeneration*, (Lilburn, GA: Fairmount Press 1988).

Heywood, John B., *Internal Combustion Engine Fundamentals*, (McGraw-Hill Book Company, Inc., 1988).

IEEE, *Recommended Practice for Emergency and Standby power Systems for Industrial and Commercial Applications*. (New York: John Wiley, 1972)

345

IEEE, *Recommended Practice for Electric Systems in Health Care Facilities*, (New York: IEEE, 1986).

IEEE, *Recommended Practice for Grounding of Industrial and Commercial power Systems,*(New York: IEEE, 1972).

Johnson, Gordon S. Editor, *On-Site Power Generation*, (Boca Raton, FL: ESGA 1998).

Jorgensen, Robert, *Fan Eneineering,* (Buffalo: Buffalo Forge Co. 1961).

Judge, Arthur W. *High Speed Diesel Engines,* (London: Chapman & Hall, 1943)

Kates, Edgar J., *Diesel and High-Compression Gas Engines*, (Chicago: American Technical Society, 1965).

Kent, M.E., Robert Thurston, *Kent's Mechanical Engineers' Handbook*, 11th ed. (New York: John Wiley & Sons, Inc., 1937).

Kent, M.E., Robert Thurston, *Kent's Mechanical Engineers' Handbook of Design and Shop Practice*, (New York: John Wiley & Sons, Inc., 1947).

Lindsey, Forrest R., *Pipefitters Handbook*, 3rd ed. (New York, N.Y.: Industrial Press, Inc., 1967).

McKinnon, Gordon P., Editor, *Industrial Fire Hazards Handbook*, (Boston: National Fire Protection Association, 1979)

North American Manufacturing Company, *Combustion Handbook*, (Cleveland, Ohio: The North American Manufacturing Company, 1952).

Obert, Edward F., *Internal Combustion Engines-Analysis and Practice*, 2nd ed. (Scranton, Pennsylvania: International Textbook Company, 1950).

RegO, *LP-Gas Serviceman's Manual L-545*, (Elon College, N.C.: RegO Products, 1962).

Shepherd, Harold F., *Diesel Engine Design,* (New York: John Wiley, 1935)

Shrager, A.B., B.S., Arthur M., *Elementary Metallurgy and Metallography,* 3rd ed., (New York: Dover Publications, Inc. 1969).

Smith, William Fortune, *Structure and Properties of Engineering Alloys,* (New York: McGraw-Hill Inc., 1981).

Smith, William Fortune, *Principles of Material Science and Engineering,* (New York: McGraw-Hill, Inc., 1999).

Taylor, C.F., *The Internal Combustion Engine,*(Scranton, PA: International Textbook Company, 1961).

Vallance, Alex, *Design of Machine Members,* (New York: McGraw-Hill, 1943).

Vinal, George Wood, *Storage Batteries,* (New York: John Wiley, 1955).

Wood, A.B. *A Textbook of Sound,*(London: G. Bell and Sons LTD, 1955).

SUPPLIERS:

FLYWHEEL COUPLINGS -ELASTIC

KTR Corporation
P.O. Box 9065
Michigan City, In 46361
ktr-us@ktr.com

Ringfeder Corp
(800) 245- 2580
www.ringfeder.com

INVERTERS:
Out Back Power Systems
19009 62 Ave. NE
Arlington, WA 98223
(360) 435 – 6030
www.outbackpower.com

Xantrex
8999 Nelson way
Burnaby, BC Canada, V5A 4B5
(604) 422-8595
www.xantrex.com

Out Back Retail Outlet:
Northern Arizona Wind and Sun
4091 East Huntington Drive
Flagstaff , AZ 86004
(800) 383-0195
www.Solar-Electric.Com

Xantrex Retail Outlet:
DonRowe.Com
Box 552
Monroe, OR 97456
(800) 367-3019
www.DonRowe.Com

GOVERNORS:
The Pierce Company
35 N. 8th Street
Upland, IN 46989
(765) 998 – 2712
www.ThePierceCompany.Com

Woodward Governor Company
5001 2nd Street.
Rockford, IL 61125
(815) 877-7441
www.Woodward.Com

TRANSFER SWITCHES:

Connecticut Electric Switch Company
5508 128th St.
East Puyallup, WA 98373
(866) 822-0453
www.Connecticut-Electric.com

Harbor Freight Tools
www.HarborFreight.Com

RONK
106 E. State St.
Nokomis, IL 62075
217 563-8333
www.RonkElectrical.Com

Eyelander Electric
P.O.Box 1749
Everett, WA 98206
(800) 932-8986
www.EylanderElectric.Com

Reliance Controls Corporation
2001 Young Court
Racine, WI 53404
(262) 634-6155
www.RelianceControls.Com

Midwest Electrical Testing & Maintenance
5141 North 35th Street
Milwaukee, WI 53209
(414) 931- 0992
www.SWGR.Com

347

PROPANE REGULATORS AND CARBS:

Carburetion & Turbo Systems
1897 Eagle Creek Blvd
Shakopee, MN 55379
(952) 445-3910
www.CarbTurbo.Com

US Carburetion
416 Main Street
Summersville, WV 26651
(800) 553-5608
www.PropaneGenerators.Com

Impco
3030 South Susan Street
Santa Ana, CA 92704
(714) 656-1212
www.impco.ws

GENERATORS*:

Stamford Newage America
7928 University Ave NE.
Fridley, MN 55432
(800) 367 2764
(610) 578-9233
www.Newage-AvkSeg.com

Marathon Electric
100 E Randolph St.
Wausau, WI 54402
(800) 477-6362
MarathonElectric.Com

RADIATORS:

1-800-Radiator
2080 Peachtree Industrial Suite 113
Chamblee, GA 30341
(800)-252-0333

Inexpensive radiators
delivered overnight
to your door.
Great Service!

ENGINE PROTECTION AND CONTROL:

Murphy Controls
105 Randon Dyer Drive
Rosenberg, TX 77471
(281) 633-4500
www.FWMurphy.Com

Deep Sea Electronics
3230 Williams Ave.
Rockford, IL 61101
(815) 316-8706
www.DeepSeaPLC.com

ALTERNATIVE ENERGY SUPPLIES:

EA Energy Alternatives Ltd.
5 – 4217 Glanford Ave.
Victoria, BC V8Z – 4B9 Canada
(800) 265-8898
www.EnergyAlternatives.Ca

DEEP CYCLE BATTERIES:

Rolls Battery Engineering
P.O. Box 671
Salem, MA 01970
(800) 681-9914
www.RollsBattery.Com

DEEP CYCLE BATTERIES:

US -Battery Co.
653 Industrial Park Dr.
Evans, GA 30809
(706) 868-0533
www.USBattery.Com

Trojan Battery Company
www.Trojan-Battery.Com

TOOLS AND MATERIALS:

Enco
400 Nevada Pacific Hwy.
Fernly, NV 89408
800 USE-ENCO
www.Use-Enco.Com

McMaster-Carr
6100 Fulton Industrial Blvd
Atlanta, GA 30336
(404) 346-7000
www. McMaster.Com

MSC Industrial Supply
75 Maxess Rd.
Melville, NY 11747
800 645-7270
www.MSCDirect.Com

INDEX:

ALTERNATIVE ENERGY SECRETS: PRACTICAL SOLUTIONS FOR PERSONAL ENERGY PROBLEMS. A companion text to GENERATORS AND INVERTERS that focuses on overlooked fuels and conversion of auto engines to multi-fuel (propane-CNG) operation. Extensive information regarding waste oils and their inexpensive conversion to useful fuels. Describes oil cracking, blending, filtering, and construction of centrifuges to clean waste oils and make them suitable for use in home heating or as diesel fuel. Practical "How To". Very unusual material. Nothing like it anywhere on the market. A must have for alternative energy users!

ISBN 978-0-9702203-6-3   (Available 2009) $19.95

MAKING PISTONS FOR EXPERIMENTAL AND RESTORATION ENGINES
You are no longer limited by the price and availability of replacement pistons and rings when you can make your own. Design and make pistons for new or old engines. Use inexpensive modern piston rings on your antique equipment! Learn to make all the tools and jigs needed to quickly produce top quality replacements in your own back yard and home shop. Heavily illustrated. A "must have" for antique equipment restorers!

ISBN 0-9702203-4-0   64 pages $11.95

METAL CASTING: A SAND CASTING MANUAL VOL. 1 & 2
Learn how to cast metal in sand molds using simple techniques and readily available materials. See how to make a sand mold and then how to hone your skills to produce high quality castings. Written in non-technical terms, the sand casting manuals begin by melting aluminum cans over a charcoal fire and end by casting a cylinder head. All for little cost in your own back yard! Sold in over 30 countries, Chastain's popular "Small Foundry Series" is good for both the beginner and experienced metal caster.
VOLUME I   ISBN 0-9702203-2-4   208 Pages $19.95
VOLUME II   ISBN 0-9702203-3-4   192 Pages $19.95

IRON MELTING CUPOLA FURNACES
Complete plans and operating instructions for a 10" diameter cupola that will melt 330 pounds of iron per hour when powered by a shop-vac! Can be built for little cost, mostly from scrap.
ISBN 0-9702203-0-8   128 pages $19.95

BUILD A TILTING FURNACE
Complete plans and operating instructions for a tilting furnace that easily melts 100 pounds of aluminum per hour. Melt with propane, diesel, or used motor oil!
ISBN 09702203-1-6   192 Pages $19.95

See more titles and color photos at: **StephenChastain.Com**